Evaluation of Flexible Spending Accounts for Active-Duty Service Members

BETH J. ASCH, PATRICIA K. TONG, LISA BERDIE, MICHAEL G. MATTOCK

Prepared for Office of the Secretary of Defense
Approved for public release; distribution unlimited

For more information on this publication, visit **www.rand.org/t/RRA1553-1**

Published by the RAND Corporation, Santa Monica, Calif.

Library of Congress Cataloging-in-Publication Data is available for this publication.

ISBN: 978-1-9774-0985-0

Cover: U.S. Army.

About This Report

The House Armed Services Committee Report accompanying the National Defense Authorization Act for fiscal year 2021 directed the Secretary of Defense to report on the feasibility of implementing flexible account options that allow pre-tax payment of dependent care expenses, health and dental insurance premiums, and out-of-pocket health care expenses for service members and their families (U.S. House of Representatives Report 116-617, 2020). These options allow employees to put aside a portion of their wage earnings into tax-advantaged accounts that can be used to pay for eligible health care and dependent care expenses. Employee contributions to these accounts interact in complex ways with other tax incentives, such as the Child Tax Credit, affecting whether such options are advantageous for members and their families. Administering these accounts also imposes a cost on the U.S. Department of Defense (DoD), although there are also sources of savings.

The Office of the Secretary of Defense asked the RAND National Defense Research Institute to provide analytic support to DoD as input for its report to the U.S. Congress, and this report documents the RAND Corporation team's research. The analysis summarized in this report focuses on the implications of flexible account options for active-duty service members and their families. This report should be of interest to policymakers and researchers who are concerned about programs targeting dependent care for military families, costs of health care and the quality of life for military families, and the adequacy of military compensation.

National Defense Research Institute

The research was sponsored by the Office of the Secretary of Defense and conducted within the Forces and Resources Policy Center of the RAND National Security Research Division (NSRD), which operates the National Defense Research Institute (NDRI), a federally funded research and development program sponsored by the Office of the Secretary of Defense, the Joint Staff, the Unified Combatant Commands, the Navy, the Marine Corps, the defense agencies, and the defense intelligence enterprise.

For more information on the RAND Forces and Resources Policy Center, see www.rand.org/nsrd/frp or contact the director (contact information is provided on the webpage).

The research reported here was completed in July 2022 and underwent security review with the sponsor and the Defense Office of Prepublication and Security Review before public release.

Acknowledgments

We are grateful to Don Svendsen, Assistant Director of the Office of Compensation, who served as project monitor, provided background material, and served as a tremendous resource for this project. We are also very grateful to Heidi Welch, Associate Director for Children, Youth, and Families with the Office of Military Family Readiness Policy within the Office of the Under Secretary of Defense for Personnel and Readiness, for providing data and information and answering our questions and Stacey Young in that office. We are indebted to the subject-matter experts who spoke with us about the potential costs to DoD of implementing flexible spending accounts for service members. We also gratefully acknowledge the help of Jonas Kempf who provided research programming support, and we wish to thank our RAND colleagues Molly McIntosh and Daniel Ginsberg, the Director and Associate Director, respectively, of the Forces and Policy Resources Center within the NDRI. Finally, our research benefited from the input of the two reviewers, Susan Gates at RAND and Paul Hogan of the Lewin Group. We greatly appreciated their input.

Summary

Unlike many large employers, including the federal government, the military does not offer flexible spending account (FSA) options to members of the armed services and their families. The House Armed Services Committee Report accompanying the National Defense Authorization Act for fiscal year 2021 directed the Secretary of Defense to report on the feasibility of implementing flexible account options that allow pre-tax payment of dependent care expenses, health and dental insurance premiums, and out-of-pocket health care expenses for service members and their families (U.S. House of Representatives Report 116-617, 2020). As detailed by the directive, the report should include an assessment of the tax incentives when using FSA options and the financial advantages or disadvantages to service members and their families and should include identification of any legislative or administrative barriers to implementing these options.

Contributions to either a health care FSA (HCFSA) and/or dependent care FSA (DCFSA) reduces the amount of income subject to income and payroll taxes, thereby reducing the individual's tax liability. FSAs interact with other tax incentives in the U.S. tax code, potentially reducing or even eliminating the potential tax savings to individuals with an FSA. From the standpoint of the employer, such as the U.S. Department of Defense (DoD), FSAs reduce Federal Income Contribution Act (FICA) taxes that consist of both Social Security and Medicare taxes but impart a cost in the form of administration costs and implementation or start-up costs. Thus, a priori, it is not possible to predict whether FSAs would represent a net benefit to service members and their families in terms of reduced tax liabilities or a net reduction in costs to DoD.

The Office of the Secretary of Defense (OSD) asked the RAND National Defense Research Institute to provide analytic support for its report to Congress, including an implementation plan should FSA options be implemented by DoD. This report documents that support. Our analysis focused on active-duty personnel and sought to evaluate the benefits and costs to service members of FSAs for dependent care and health care and the costs and savings to DoD. For contextual background, we drew information from policy documents and other sources on existing health care and dependent care benefits available to military members and civilians. To help understand the elements of cost and savings to DoD, we held structured discussions with subject-matter experts (SMEs) on administrative costs and we also gathered available inputs on administrative costs. We used publicly available tax simulation programs to evaluate the tax benefits and costs to service members of FSAs for dependent care and health care and compute the aggregate net change in tax liabilities to service members and the net costs and savings to DoD under a variety of assumptions regarding the share of members who might participate in the FSA options. Finally, we drew together the results of the analysis to assess the feasibility of providing FSA options to service members and discuss potential legislative changes that could improve the net benefit of FSAs to members and their families. We note that our tax-accounting approach does not consider how FSAs might change the members' incentives to consume health care or dependent care or issues related to the efficiency or social welfare implications of providing a health and dependent care benefit through the U.S. tax code.

Key Findings

Many Members Would Have Few or Even No Eligible FSA Expenses Under Current Law

For members to take advantage of an FSA, they must have eligible expenses, usually in the form of dependent care expenses and out-of-pocket health care expenses for themselves or their family members. In the case of health care, most members would have few or no eligible out-of-pocket medical care costs associated with TRICARE because 82 percent of family members were enrolled in TRICARE Prime and incurred an average of $100 annually in out-of-pocket health care costs between 2018 and 2020. Meanwhile, 17 percent of family members enrolled in TRICARE Select and incurred higher costs, an average of about $500. HCFSAs also cover health-related expenses outside TRICARE, including over-the-counter medication, eyeglasses and contact lenses, and orthodontia expenses. Estimates for these expenses for the active-duty population are unavailable, but national estimates suggest that out-of-pocket expenses outside health insurance are about $400, although the U.S. population differs demographically from the military population and out-of-pocket expenses could also differ. We note that orthodontia expenses specifically have an estimated average cost of $3,100[1] in 2021 dollars among those incurring these expenses (Hung et al., 2021).

Dependent Care Expenses

In the case of dependent care, because of the way in which DoD provides child care benefits, an unknown share—and, potentially, a large share—of military families may not have eligible dependent care expenses. Using data from the 2019 Survey of Active Duty Spouses conducted by DoD (Office of People Analytics, DoD, 2020), we estimate that 65 percent of military spouses are working and have sufficient earnings to qualify for a DCFSA and have sufficiently young children to potentially benefit from a DCFSA.[2] However, expenses for

[1] Hung et al. (2021) estimates that average orthodontic expenses for those with any expenses were about $2,800 in 2016 using Medical Expenditure Panel Survey data. Inflating this amount to 2021 dollars suggests that average orthodontic expenses were about $3,100 in 2021.

[2] The 65-percent figure is computed as 42 percent divided by 64 percent. The 42-percent figure reflects all spouses with young children who use child care, not working spouses. To qualify for a DCFSA, the working spouse must have earnings at least equal to the DCFSA contributions. According to the 2019 spouse survey, 64 percent of military spouses are in the labor force. Note that we assume the statistics reported in the 2019 spouse survey apply to the DCFSA eligible population even though the 2019 spouse survey asks about child care for children age 13 and younger and eligible expenses for DCFSAs must be for children *younger* than age 13.

off-base child care subsidized by DoD are unlikely to be FSA eligible. The reason is that the subsidy offsets the eligible expenses for a DCFSA dollar for dollar; any combination of child care subsidy and DCFSA contributions cannot exceed the maximum DCFSA contribution limit of $5,000. That is, expenses that would be eligible for DCFSA contributions and expenses that are subsidized are not *stackable*.[3] According to the 2019 Survey of Active Duty Spouses, 61 percent of spouses who use child care report using child care that is provided off base by a civilian provider. Many military families seek off-base care because care at on-base child development center (CDC) facilities is *supply constrained*, meaning that there are more members seeking care for their children at these centers than there are spaces for children, and CDCs have wait lists (Kamarck, 2020). We have no information on what share of these 61 percent of spouses accesses fee-assisted or subsidized care and therefore would be unlikely to have eligible DCFSA expenses. It is likely that some of this off-base care is unsubsidized (implying that expenses would be FSA eligible), as we discuss in the next subsection, but more information is needed on the extent.

FSAs Would Confer a Tax Savings for Many But Not All Members with Eligible Expenses

For those military families who do have eligible expenses, we estimate that the HCFSA option would confer a savings but that the DCFSA would not do so in all cases. Contributions to an FSA reduce the amount of income subject to taxation, thereby reducing the amount of income tax and payroll taxes (the employee's contribution to Social Security and Medicare taxes) that a member would pay, suggesting that FSAs would produce a tax savings for service members. However, FSA contributions interact with other components of the U.S. tax code, specifically the Child Tax Credit, the Child and Dependent Care Tax Credit (CDCTC), and the Earned Income Tax Credit (EITC), sometimes in ways that are not necessarily favorable on net to the taxpayer. These credits have phase-in, plateau, and phase-out points as income changes—and, by reducing taxable income, FSA contributions affect how much of these credits can be claimed and the tax liability of the taxpayer. Consequently, on net, the tax liability for service members does not necessarily decrease in all cases.

We estimate the total change in tax liability for married and unmarried service members under different assumptions about the number of children, whether spouses are working, and FSA contribution amounts, and we consider results by paygrade and by total family income (TFI) (before FSA contributions). Considering results by TFI allows us to consider cases where members might earn special and incentive pay (raising family income) or spouses

[3] If expenses were *stackable*, out-of-pocket expenses paid by the member for child care would not be offset by the amount of the subsidy. For example, suppose a member's total child care fee is $20,000 annually, with the member facing $5,000 in out-of-pocket expenses and the DoD subsidy to the community-based provider is $15,000. Because $15,000 exceeds $5,000, and the subsidy offsets DCFSA eligible expenses dollar for dollar, the $5,000 out-of-pocket expenses would not be DCFSA eligible. If, instead, expenses are stackable, the $5,000 would be eligible.

have higher than average earnings. (We estimate that average spouse earnings were $25,000 in 2020).

In the case of dependent care, we estimate that the total change in tax liability from contributing $5,000 to a DCFSA would not always be advantageous to the service member. That is, we find cases where the tax liability remains unchanged or increases slightly.[4] Figure S.1 illustrates the change in total tax liability of contributing $5,000 to a DCFSA for a married service member in 2020. The figure shows the complex, nonlinear nature of the U.S. tax code with the tax change rising and falling at different earnings levels as certain credits phase in and out and as tax brackets change. In general, for lower levels of earnings, the tax benefit from contributing to a DCFSA is driven by increases in the amount of EITC that can be claimed, and to a smaller extent, increases in the amount of the refundable Child Tax Credit. As the EITC phases out at higher earnings levels, the tax benefit from contributing $5,000 to a DCFSA decreases. Once the EITC phases out (i.e., earnings exceed the means tested threshold to claim the EITC), the tax benefit from contributing to a DCFSA plateaus until earnings reach a level where the service member moves up to the next tax bracket and the reduction in income tax (before credits) increases. Additional nonlinearities in the tax code cause the tax benefit from contributing to a DCFSA to increase and decrease at higher earnings levels.

FIGURE S.1

Change in Taxes by Gross Earnings from Contributing $5,000 to DCFSA, Married Service Members, 2020

NOTES: Gross earnings are equal to family earnings without the FSA contribution deducted. The x-axis in the figure shows gross family earnings, meaning these are earnings without the $5,000 reduction from the DCFSA contribution. For married service members, we assume that the military spouse earns at least $5,000, so the couple qualifies for the maximum DCFSA contribution. We used publicly available tax simulation programs to estimate the change in federal income tax from contributing $5,000 to a DCFSA and add in the estimated payroll tax reduction of $382.50 to calculate the total tax change.

---

[4] We note that the figures are estimates of hypothetical tax increases. Because participation in an FSA option would be voluntary, those who expect a tax increase would be unlikely to participate, so their tax change due to participation would be zero in reality.

The results in Figure S.1 show that over the range of $59,000 to $105,000 of TFI, a married service member with two children and $5,000 of eligible DCFSA expenses would not benefit from having the FSA; the change in total tax liability is virtually zero (or, more precisely equal to +$17.50). This occurs because of a loss in the CDCTC, which exceeds the reduction in tax before credits and payroll taxes.

We estimate that virtually all members with eligible health expenses would experience a tax savings from contributing to an HCFSA regardless of marital status and number of children with the tax savings increasing when HCFSA contributions are larger. Figure S.2 shows the change in total taxes from contributing $500 to an HCFSA for a married member with a working spouse. Although all members are estimated to experience a savings in Figure S.2, the amount varies with income because of the phase-in and phase-out of different credits at different income levels and because the tax brackets vary with income.

Across Service Members, FSAs Would Confer a Net Tax Benefit

Ideally, to estimate the total net change in tax savings or costs across service members, we would compute the share of members who would likely participate in the HCFSA and DCFSA options and then compute the cost or savings for those members and for DoD. Because we lack sufficient data to estimate the HCFSA and DCFSA participate rates, we compute savings and costs under alternative participation rate assumptions and then use available, albeit incomplete, information on which participation rates seem the most likely.

For the DCFSA option, we use (1) Defense Manpower Data Center data for the number of service members on active duty in 2020 and the number with children under age 13 and (2)

FIGURE S.2

Change in Taxes by Gross Earnings from Contributing $500 to HCFSA, Married Service Members, Working Spouse, 2020

NOTE: Gross earnings are equal to family earnings without the FSA contribution deducted.

2019 Survey of Active Duty Spouses on the share of service members with a working spouse to estimate the total tax savings to service members contributing $5,000 to a DCFSA under different DCFSA program participation rates. We estimate an annual aggregate tax savings of $30.7 million if 15 percent participate, $51.2 million if 25 percent participate, $102.4 million if 50 percent participate, and $204.9 million if 100 percent participate. A participation rate of between 25 percent to 50 percent may be more realistic than a 15-percent or 100-percent participation rate. As discussed earlier, 64 percent of employed spouses report using child care, an upper estimate of the rate of participation in a DCFSA. However, the 64-percent figure is too high as an estimate of a participation rate because not all of these military families with an employed spouse would have eligible DCFSA expenses given that some families receive subsidized off-base child care that offsets the tax benefit of a DCFSA dollar for dollar. The more realistic estimate is somewhere above 24 percent but below 64 percent. The 24 percent corresponds to our estimate of working spouses using child care who would use on-base child care.[5] But some families with working spouses use off-base unsubsidized care that would be eligible for a DCFSA, so the participation rate would be higher than 24 percent, but we do not have an estimate of how much higher.[6] Given this context, a participation rate between 25 percent to 50 percent may be more realistic leading to an estimated aggregate benefit of $51.2 million to $102.4 million.

In the case of an HCFSA, we assume that all active-duty service members are eligible to participate in an HCFSA. We estimate an annual aggregate tax savings of $19.8 million assuming 15 percent of active members participate in the HCFSA and contribute $500 annually. That figure increases to $33.1 million assuming a 25-percent participation rate to $66.1 million assuming 50 percent participate to $132.3 million assuming a 100-percent participation rate. Given that 17 percent of active-duty families were covered by TRICARE Select and averaged $500 annual out-of-pocket expenses, the actual participation rate may be closer to the 25-percent figure than the 50-percent figure. On the other hand, all members could have other eligible out-of-pocket expenses outside TRICARE, equal to about $400 for the civilian population, so participation could be higher.

[5] Of the 64 percent of working spouses who report using child care in the 2019 survey, 37 percent (or 24 percent of working spouses using child care) would have eligible expenses because they use on-base child care.

[6] It is likely that some families do use unsubsidized off-base care given the differences between the 2019 DoD statistics, discussed in Chapter Three, on the number of children enrolled annually in military-supported child care and responses to the 2019 spouse survey indicating that 73 percent of spouses using child care report using off-base care. The DoD statistics imply that only 9 percent to 16 percent of enrolled children are in off-base fee-assisted care. But, if all 73 percent of spouses using off-base care received only fee-assisted care, then the 9-to-16-percent figure is too low. Thus, it is likely that some unknown share of the 73 percent is receiving unsubsidized care. Future surveys of active-duty spouses should ask more detail about whether these spouses are receiving fee-assisted care and would be unlikely to have eligible DCFSA expenses.

FSAs Would Confer a Cost Savings to DoD but Could Result in Large, Albeit Unknown Implementation Costs

We also estimated the aggregate cost to DoD's accounting of administering FSA options and the savings to payroll taxes from reduced Social Security and Medicare taxes. We used information from the U.S. Office of Personnel Management (OPM) (OPM, 2018; OPM, 2020; and OPM 2021) on the cost of administering the FSA options for federal employees, noting that administration costs are lower after the first year when more reserve funds are built up to handle overpayments and other situations. Table S.1 shows the estimated cost implications under different assumptions about participation rates assuming members contribute $5,000 to a DCFSA (top panel) or contribute $500 annually to an HCFSA (bottom panel). Assuming a 25-percent participation rate for each program, we estimate that DoD would save $6.0 million annually for the HCFSA program and $31.6 million for the DCFSA program, after the first year. These figures exclude implementation costs and ongoing overhead costs, such as the cost of training personnel about the FSA options.

We held discussions with SMEs to better understand these implementation and ongoing overhead costs. The experts identified costs associated with developing the processes and contractual arrangements required to administer the FSA plans and the costs of providing financial literacy training to members. In addition, the pay systems would need to be modified to accommodate the FSA options. They noted that using off-the-shelf training and OPM to assist in plan administration would help reduce the first two costs, but the costs of adapting the pay systems could potentially be high if implementation involved adapting both the legacy pay systems and the forthcoming service-specific systems, ranging up to $28 million or more. The services are each undertaking the development of an integrated pay and personnel system, and each is expected to roll out these systems in the coming years, although on different schedules. Because of this change in systems and the different service schedules, the legacy pay systems would need to be modified, not just the systems under development. Implementation costs would be lower if DoD only implemented the FSA options in the new

TABLE S.1
DCFSA and HCFSA Savings to DoD

	FSA Participation Rate			
	15%	25%	50%	100%
HCFSA, assuming $500 contribution per member				
Total Savings Year 1	–$181,000	–$302,000	–$604,000	–$1,207,000
Total Savings Year 2+	$3,600,000	$6,000,000	$12,000,000	$24,100,000
DCFSA, assuming $5,000 contribution per member				
Total Savings Year 1	$17,900,000	$29,800,000	$59,500,000	$119,100,000
Total Savings Year 2+	$19,000,000	$31,600,000	$63,300,000	$126,500,000

NOTE: Cost estimates include annual contractor administrative costs and payroll tax savings. Cost estimates do not include implementation costs or ongoing overhead costs. A negative DoD savings means that DoD incurs a cost.

systems, thereby avoiding the necessity of changing the legacy pay system. But, even focusing on only the new systems, several of the experts observed that adapting or creating the pay systems needed to support FSAs could prove to be a substantial challenge, both in terms of cost and time, and could result in delays in the roll out of the new systems.

Legislative Changes Could Expand the Benefit of an FSA to Service Members

Our analysis estimates the costs and benefits to service members and their families of FSA options under current law. However, changes in current law could expand the value of an FSA to members. First, current regulations do not allow for health FSA funds to be applied to health and dental insurance premiums. Thus, enabling legislation would need to be passed to allow for payment of insurance premiums from FSAs for military personnel. Active members and their families have no premium or enrollment fee for TRICARE Prime or Select, but families pay a premium for dental care equal to $11.65 and $30.28 per month in 2022, respectively, for single members and members with families. Second, members who receive subsidized off-base child care would be unlikely to benefit from a DCFSA, because contributions to an FSA are offset by the subsidy. This issue could be addressed by eliminating the offset for the employer subsidy of child care. However, eliminating the offset would require legislation to make an employer child care subsidy and eligible DCFSA expenses stackable. Allowing them to be stackable would likely increase participation in the DCFSA. Finally, FSAs for service members could be implemented faster—perhaps up a year faster—if enabling legislation to support special features, such as payment of insurance premiums, was not required.

Wrap-Up

Introducing an FSA option could impart a tax benefit for some members and their families and save personnel costs for DoD, although the start-up costs of updating the legacy and new pay systems could be substantial. However, many military families would not have eligible DCFSA expenses, and families could have few eligible HCFSA expenses under current law, unless they use the HCFSA to cover expenses outside TRICARE, such as for over-the-counter medication and supplies. Changes in law could expand the value of FSA options to service members.

Contents

Figures and Tables

Figures

Tables

CHAPTER ONE

Introduction

The report of the House Armed Services Committee accompanying the National Defense Authorization Act (NDAA) of 2021 directed the Secretary of Defense to submit a report on the feasibility of implementing flexible spending account (FSA) options that allow pre-tax payment of dependent care expenses, health and dental insurance premiums, and out-of-pocket health care expenses for members of the uniformed services and their family members (U.S. House of Representatives Report 116-617, 2020). As detailed by the directive, the report should include (1) an assessment of the tax incentives when using FSA options and the financial advantages or disadvantages to service members and their families and (2) identification of any legislative or administrative barriers to implementing these options. Health care FSAs (HCFSAs) and dependent care FSAs (DCFSAs) are an employee benefit that allows workers to put aside a portion of their wage earnings into tax-advantaged accounts that can be used to pay for eligible health care and dependent care expenses. Employee contributions to FSAs are not subject to income or payroll taxes, such as Social Security and Medicare taxes (also known as Federal Income Contribution Act [FICA] taxes). Thus, the benefit to employees is the tax savings from reducing their gross income by the amount contributed to the FSA(s). Because employee contributions reduce the wage base on which FICA taxes are calculated, the amount that the employer must contribute to Social Security and Medicare—the employer portion of the payroll tax—is also reduced. Federal government agencies, including the U.S. Department of Defense (DoD) provide civilian employees with access to FSAs and certain private and public sector employers offer FSAs to their employees.

FSAs are not currently available to uniformed service members in DoD. Although FSAs could confer a benefit to service members, they also have requirements. Employees must actively elect annually to participate, and their elections are irrevocable unless the employee has a qualifying life event, such as a marriage or birth. Also, accounts have a "use-or-lose" feature, meaning that the employee loses any contributions remaining that were not spent on eligible expenses within the required timeframe. From the standpoint of DoD, its portion of FICA taxes is lower than it would be in the absence of the FSA options but administering FSA accounts would involve an ongoing maintenance cost and set-up costs. DoD would also face administrative costs to reconfigure its pay systems and it would need to provide financial education to train the force about this new benefit.

Because FSAs could be attractive to both members and DoD, but also involve requirements and costs, Congress required a report on the potential benefits and costs of providing service members with the option to contribute to pre-tax FSAs for dependent care and health care expenses. The report should also identify any legislative or administrative barriers to achieving the implementation of such an option. For example, the congressional directive for a DoD study indicated that FSAs for service members would cover premiums for health and dental insurance, and these are not eligible expenses employees under current law governing FSAs.

The Office of the Secretary of Defense (OSD) asked the RAND National Defense Research Institute to provide analytic support to DoD as input for its report to Congress, including an implementation plan should FSA options be implemented by DoD. This report documents that support. Our analysis relied on a mixed-method approach and focused on active-duty personnel. We drew information from policy documents and other sources on existing health care and dependent care benefits available to military members and to civilians, held structured discussions with subject-matter experts (SMEs) on administrative costs and gathered available inputs on administrative costs, and used publicly available tax simulation programs together with key data inputs to evaluate the tax benefits and costs to service members of FSAs for dependent care and health care. In the process, we identify subsets of military personnel who could benefit from FSAs and others who might not have sufficient eligible expenses to make them worthwhile. Furthermore, given the directive's requirement to consider legislative and administrative barriers, we also estimate the ongoing administrative costs and implementation costs to DoD, consider possible legislative changes, and use this information to develop an implementation plan should DoD move forward with the FSA options. The results of our analysis are presented in this document.

Caveats

Our methodology relies on a tax-accounting approach that recognizes that FSA options have tax incentive implications that interact with other tax credits and elements of the tax code. This approach responds to Congress' directive, but we note several important caveats. First, we do not provide a full accounting of all taxation implications. Although we consider the tax advantages and disadvantages to service members, their families, and DoD, we do not consider the implications for general taxpayers. Insofar as FSA options confer a tax savings to DoD and therefore the government, we do not consider the implications of foregone government tax revenues.

Second, we do not fully consider the variety of social welfare implications of FSA options for military personnel, particularly those related to the economic efficiency of these accounts. Tax-preferred accounts potentially drive a wedge between the marginal value of tax-favored expenditures and of nonpreferred expenditures; although it may be the case that a dollar of expenditure on an item that is eligible for tax preference is worth the same at the margin as an unconstrained dollar of expenditure (i.e., a dollar that would be spent on an item if tax preference were not available), that is not necessarily so. For example, FSAs have a use-it-or-lose it feature whereby participants lose the funds in the FSA account if they are not spent by a certain date, such as the end of the calendar year. Such a provision can create of flurry of expenditures in December as participants attempt to use up their FSA funds, but those expenditures may not be worth as much as spending those dollars in an unconstrained manner. If these tax-preferred expenses are worth less, then at the margin, the net benefit to the member compared with not having an FSA would be less than we report. More generally, FSA options might induce changes in members' behavior and specifically could induce members to consume more health care or dependent care than they would in the absence of tax preference, as discussed by Lo Sasso et al. (2010) in the context of health care accounts. Although the increase in member consumption of health care could be deemed a desirable outcome on the part of policymakers, it could also be inefficient; our analysis does not attempt to measure any induced changes in member behavior or their efficiency or cost implications.

Another behavior change that we do not consider is the potential change in the type of members recruited and retained or the overall number recruited and retained. FSAs could increase the attractiveness of military service for married members with dependents or those intending to marry and acquiring dependents, although any effect is likely to be small relative to the effects on recruiting and retention of other policies, such as a pay raise. Such behavioral changes could have implications for efficiency and DoD costs. Should FSAs produce such an effect, DoD costs could potentially increase to the extent that health care, housing, and the other costs are higher for members with dependents than for members without them. Furthermore, although an increase in recruiting and retention would likely be welcomed by the armed services in terms of meeting their force size objectives, using such policies as basic pay and bonuses that are neutral with respect to marital and dependent status would likely be more efficient than using FSAs to achieve these objectives.

Organization of This Report

Our report is organized as follows. Chapter Two reviews the landscape of health care and dependent care benefits available to military members and civilians. Chapter Three provides details on how, conceptually, an FSA option could benefit service members and their families. In Chapter Four, we review the data and methods we use to compute benefits and costs to service members and their families, and present our estimates. In Chapter Five, we estimate FSA costs and savings to DoD and the total savings to member, and we discuss what we learned from the stakeholder discussions regarding administrative costs. In Chapter Six, we provide an implementation plan and identify areas of uncertainty and risk and factors affecting cost. Finally, in Chapter Seven, we draw conclusions about the advisability and feasibility of an FSA option for active-duty service members.

CHAPTER TWO

Landscape of Dependent Care and Health Care Benefits Relevant to the Discussion of FSA Options for Military Personnel

In assessing the potential benefit of FSAs to active military members, it is important to understand the broader context of benefits available to military personnel. As discussed in Chapter One, FSAs provide savings by excluding taxable income, meaning beneficiaries pay taxes on less income than they otherwise would. Because these arrangements provide savings by reducing beneficiaries' tax liabilities, we review FSAs and other tax-based benefits in health care and dependent care from two sources. The first is the set of benefits available to the broader U.S. taxpayer population and that may be available to military personnel and/or their spouses. The second is the set of military-sponsored health care and dependent care benefits that military members might access. Both sources determine whether and how service members would experience a tax benefit from contributing to an FSA.

The Landscape of Health Care Benefits

We first review health care tax benefits that are available to U.S. taxpayers more broadly and that military personnel or their spouses may be able to access, and then we review military-specific health benefits.

Health Care Tax Benefits

A variety of tax programs are aimed at reducing the cost of medical expenses to U.S. taxpayers. First, several tax-advantaged accounts that enable taxpayers to put nontaxable income toward their medical expenses; an HCFSA is one such account. We review three types of tax-advantaged accounts in the health care domain: an FSA, health savings account (HSA), and health reimbursement arrangement (HRA). Second, several health-related tax deductions and credits that individual taxpayers can claim if relevant to their individual situation are also potentially available to American taxpayers, including uniformed service members. Table 2.1 summarizes the key features of three types of tax-advantaged health care accounts, and whether, if implemented, they could be relevant to service members.

TABLE 2.1
Comparison of Tax-Advantaged Health Care Accounts

	HCFSA	HSA	HRA
Structure	Contributions of pre-tax dollars to pay for qualified medical expenses	Contributions of pre-tax dollars to pay for qualified medical expenses	Contributions of pre-tax dollars to reimburse employee's qualified medical expenses
Account ownership	Employer	Employee	Employer
Portability after job separation	No	Yes	No
Who contributes	Employee and/or employer	Employee and/or employer	Employer
Health plan requirements	Traditional health plan (not a high-deductible health plan)	High-deductible health plan	Either traditional or high-deductible health plan (must comply with Affordable Care Act)
Fund expiration	Funds expire; grace periods and rollover limits vary (up to $570 in 2022)	Funds never expire	Funds expire
Can be used to pay insurance premiums	No	No	Often, yes (employers determine structure of reimbursement)
Potential relevance to service members	If offered, service members could use	Unlikely for service members to benefit because TRICARE does not offer a qualifying high-deductible health plan	If offered, service members could use

SOURCE: IRS (January 6, 2022).

An HCFSA is a tax-advantaged account that enables the account holder to pay for their medical expenses with pre-tax dollars. These accounts must be established through an employer who provides this benefit. Once the account is created, employees and/or employers can set aside up to a limit that is set by the Internal Revenue Service (IRS) and that is annually indexed for inflation. In 2022, the contribution limit is $2,850 annually (IRS, 2022e, p. 4). These contributions are excluded from an employee's taxable income. Employer plans can allow up to $570 of unused contributions to be rolled over to the next year, or extended during a grace period, but otherwise funds not used by the end of the plan year are forfeited (IRS, 2021b, p. 14) This is known as a *use-it-or-lose it provision*. As a result, as noted in Chapter One, the value of the benefit may be less than the contributed amount if the employee contributes more than they are able to use within the plan year and grace period. The accounts can be used to pay for qualified out-of-pocket expenses including medical expenses, co-pays, co-insurance, and deductibles. The accounts can also be used for dental and vision expenses, prescription, and over-the-counter treatments. Under current statute, HCFSA contributions cannot be used for insurance premiums (IRS, 2022a, p. 17). HCFSAs are covered by the uniform coverage rule, which requires that the maximum amount of reimbursement be available to the employee at all times during the period of coverage (IRS, 2013, p. 4). Effectively, HCFSA funds are available to cover eligible expenses at the very beginning of the coverage year, even though the employee has not yet made the full year's pre-tax contributions from their salary.

Similarly, an HSA reduces the account holder's tax liability. Like an HCFSA, an HSA is a tax-exempt account where the employee or employer can contribute funds that are excluded from the employee's income for the purposes of computing income tax. These funds can be used for nearly identical purposes to HCFSA funds with a few exceptions around allowable insurance premium-related expenses. Unlike the HCFSA, which is sponsored and held by the employer, an HSA is held by the employee. This means that the account is portable. It is also not subject to the same use-it-or-lose it restrictions that govern HCFSA funds, and the value of the expenditure is more likely to equal the contributed amount. This account type is only available to those with a high-deductible health plan in which the enrollee's out-of-pocket expenses and deductibles are expected to be sizable—and, therefore, the contribution limits are higher, and there is no rollover limit. Notably, TRICARE, the military health insurance provider, does not offer a high-deductible health plan.[1] This account type is unlikely to be relevant to active military members who currently have access to other lower-cost plan types.

A final account type is an HRA which is an account funded solely by an employer. The funds are distributed tax-free to reimburse employees for qualified medical expenses. Unlike HCFSAs and HSAs, HRA funds can be used for insurance premiums (IRS, 2022a, p. 18). This account is not currently established as a benefit for military members. Civilian federal employees are only offered this account if they are enrolled in a high-deductible health plan but are ineligible for an HSA, another circumstance that has no current corollary in the active military community.

Tax-advantaged accounts are not the only way that taxpayers can leverage tax benefits to help cover the costs of health care. There are also health care-related deductions and credits that individual taxpayers can claim. For example, taxpayers can choose to itemize deductions rather than taking a standard deduction on their taxes (IRS, 2022b). If they do so, they can deduct medical and dental expenses incurred during the tax year that exceed 7.5 percent of the taxpayer's adjusted gross income (AGI). That is to say that, if a taxpayer's AGI was $50,000, they could deduct qualified expenses beyond the first $3,750 of medical expenses paid. The list of expenses that qualify as medical-related expenses are broader than the list of expenses covered by the tax-advantaged accounts described earlier because eligible deduct-

[1] In an unpublished review of high-deductible/high co-pay health plans for military members, Paul Hogan of the Lewin Group considered a proposal that would involve offering military members such a health plan under TRICARE that would be coupled with an HSA to allow members to cover higher out-of-pocket costs, arguing that such of proposal could reduce military health care costs while holding members financially harmless.

ible expenses also include transportation expenses to access health care and insurance premiums not paid by the taxpayer's employer. Military members can already take advantage of itemizing their deductions and itemizing qualified medical expenses. Participation in a tax-advantaged account would not affect members' ability to itemize their taxes and benefit from doing so. However, as is the case for all taxpayers, any reimbursements that members received from the tax-advantaged account could not then also be itemized. For example, if a taxpayer uses an HCFSA to pay for orthodontia, the taxpayer cannot also then itemize that expense on a tax return.

Military Health Care Benefits for Active-Duty Service Members and Families

Active-duty service members and their families access health care through the Military Health System, specifically through the TRICARE health benefit program that provides benefits to members and their families. The most relevant for our analysis are the two largest health plans: TRICARE Prime and TRICARE Select.[2] This section focuses on the structure of TRICARE benefit coverage and anticipated costs that would affect the utility of HCFSA for military members (see Table 2.2).

TRICARE Prime is a health plan most comparable with a health maintenance organization (HMO) plan in the civilian sector. Under TRICARE Prime, the service member or family member is assigned a primary care manager (PCM) who serves as a primary contact, providing direct care and managing relevant referrals for care from other providers. Under TRICARE Prime, costs incurred by active members and their families are quite small. Active-duty military members and their families have no enrollment fee or annual deductible. Furthermore, care that is provided or referred by a member's PCM has co-payments and cost-shares of $0 and 0 percent, respectively. Active members do not have co-payments or cost-shares for care, even when accessed without a referral (TRICARE, 2022e). Family members of active-duty service members, however, are responsible for co-payments when they receive care without a referral. TRICARE Prime includes a catastrophic cap that limits the full amount a member and/or family would be responsible to pay for care.

TRICARE Select is a health plan most comparable with a preferred provider organization plan (PPO) in the civilian sector. Under TRICARE Select, a member or family-member accesses both in-network and out-of-network care providers without a referral. There is no enrollment fee or premium, but there is an annual deductible, which varies based on whether the member is enrolled in an individual or family plan, and the service member's grade. Co-payments and cost-sharing amounts vary by the type of care, and whether providers are

---

[2] Other TRICARE health plans—such as TRICARE Prime Remote, TRICARE Prime Remote for Active Duty Family Members, TRICARE for LIFE, and plans for beneficiaries, including former dependent children, reservists, and retirees—were reviewed for this report. The details of these plans and coverage are not included in text for simplicity because they are targeted to specific communities or situations, and are often structured similarly to the overarching Prime and Select health plans.

TABLE 2.2

Out-of-Pocket Health Care Costs for Active-Duty Service Members and Families

Out-of-Pocket Cost	TRICARE Prime Group A	TRICARE Prime Group B	TRICARE Select Group A	TRICARE Select Group B
Annual enrollment fee				
Individual	$0	$0	$0	$0
Family	$0	$0	$0	$0
Annual deductible				
E1–E4 individual	$0	$0	$50	$56
E1–E4 family	$0	$0	$100	$112
E5 and above, individual	$0	$0	$150	$168
E5 and above, family	$0	$0	$300	$336
Annual catastrophic cap	$1,000	$1,120	$1,000	$1,120
Preventative care visit	$0	$0	$0	$0
Primary care	$0	$0	$24 in network 20% out of network	$16 in network 20% out of network
Specialty care	$0	$0	$38 in network 20% out of network	$28 in network 20% out of network
Emergency room visit	$0	$0	$99 in network 20% out of network	$44 in network 20% out of network
Urgent care center visit	$0	$0	$24 in network 20% out of network	$22 in network 20% out of network
Ambulatory surgery	$0	$0	$25 in or out of network	$28 in network 20% out of network
Ambulance, outpatient (ground)	$0	$0	$74 in network 20% out of network	$16 in network 20% out of network
Ambulance, outpatient (air)	$0	$0	20% in or out of network	20% in or out of network
Durable medical equipment	$0	$0	15% in network 20% out of network	10% in network 20% out of network
Inpatient admission	$0	$0	$20.75/day; $25 minimum per admission	$67/admission in network; 20% out of network
Inpatient skilled nursing facility	$0	$0	$20.75/day; $25 minimum per admission	$20 per day in network; $56 per day out of network

SOURCE: Military Health System (undated).

in or out of network. Similar with TRICARE Prime, members in TRICARE Select have a catastrophic cap, setting the outer bounds of health care costs incurred (TRICARE, 2022d).

These health plans determine how members are accessing care, and the extent to which they choose to pay out-of-pocket expenses for that care. Both members and families enrolled in Prime and Select also incur costs for prescription medicines. When generic and brand name prescriptions are accessed through a military pharmacy, co-pays are $0. These payments increase when the medicines are accessed through home delivery or an external pharmacy, and they vary by whether they are generic, brand name, or are not included in the formulary. Over-the-counter medicines are not covered under TRICARE health plans. The vast majority of expenses are covered for active-duty members under either TRICARE Prime or Select. However, family members of active-duty service members are responsible for more expenses, thereby increasing their out-of-pocket costs.

Active-duty service members also access dental care through the Military Health System. For active members within the United States or U.S. territories, dental care is provided through military dental clinics through the Active Duty Dental Program. Service members can also receive care through civilian providers if they are referred through a military dental program, or if they live in a remote location without access to a military dental clinic. The Active Duty Dental Program has no out-of-pocket costs for active members (TRICARE, 2021).

Family members of active members are also able to receive dental coverage through the TRICARE Dental Program. A voluntary program, family members can choose to enroll in the coverage. Enrollees pay both monthly premiums and cost-shares for dental services. The costs vary by whether an individual or family is enrolling in the plan, whether the beneficiary is within or outside the continental United States (CONUS), and the paygrade of the service member. These costs are shown in Table 2.3. For coverage of a single individual, annual costs are $139.80 for enrollment premiums, plus cost shares for care beyond diagnostic and preventative care. For coverage for a family, annual costs are $363.36 for enrollment premiums, plus cost shares for care beyond diagnostic and preventative care.

The Landscape of Child and Dependent Care Benefits

As with health care, we first review child care tax benefits that are available to the U.S. taxpayers more broadly and that military personnel or their spouses may be able to access, and then we review military-specific child care benefits. Many of the tax benefits reviewed changed significantly for tax year 2021 through the American Rescue Plan. These changes were not permanent, so the information included is relevant to tax year 2020 and 2022. Information about the 2021 tax benefits is noted in footnotes.

Child and Dependent Care Tax Benefits

Tax expenditures and programs enable taxpayers to pay for child and dependent care with pre-tax dollars. To best understand how these programs function, we reviewed the benefit

TABLE 2.3

TRICARE Dental Program Costs for Family Members of Active Service Members, 2022–2023

	CONUS		Outside CONUS
	Paygrades E-1 to E-4	Paygrades E-5 and Above	All Active-Duty Families
Monthly enrollment premiums	Single: $11.65 Family: $30.28	Single: $11.65 Family: $30.28	Single: $11.65 Family: $30.28
Plan maximum (TRICARE Dental covers up to the maximum)	Nonorthodontic services: $1,500 per person/year Orthodontic services: $1,750 per person/ lifetime	Nonorthodontic services: $1,500 per person/year Orthodontic services: $1,750 per person/ lifetime	Nonorthodontic services: $1,500 per person/year Orthodontic services: $1,750 per person/ lifetime
Cost shares for dental care			
Diagnostic, preventative care, and sealants	0%	0%	0%
Consultation/office visit	20%	20%	0%
Post-surgical services and basic restorative services	20%	20%	0%
Endodontic and periodontic care	30%	40%	0%
Oral surgery	30%	40%	0%
General anesthesia	40%	40%	0%
Intravenous sedation	50%	50%	0%
Miscellaneous services (occlusal guard, athletic mouth guard)	50%	50%	0%
Other restorative care, implant services, and prosthodontic care	50%	50%	50%
Orthodontic (coverage limits apply)	50%	50%	50%

SOURCES: TRICARE, 2022c; TRICARE, 2022d; TRICARE, 2022e.

purpose and structure, and eligibility requirements and restrictions. In Chapters Three and Four, we investigate how these tax credits and programs interact and influence the tax benefit of a DCFSA.

A DCFSA is a tax-advantaged account that can be used by taxpayers to pay for eligible dependent and child care expenses with pre-tax dollars. DCFSAs can be used to pay for care for children under the age of 13, or for care for a dependent spouse or relative who is unable to care for themselves, and lives in the home. Eligible expenses include care that enable the taxpayers to work or look for work; ineligible expenses include babysitting, drop-in care, school tuition other than preschool, and meals. Contributions of up to $5,000 annually for single or married taxpayers, and up to $2,500 annually for married taxpayers filing separately, to the DCFSA are excluded from the taxpayer's gross income, reducing their tax liability.[3] Contributions cannot exceed the earned income of the taxpayer, or the taxpayer's spouse if married. This means that to contribute the maximum of $5,000 to a DCFSA, either a single taxpayer or both spouses of a married couple would need to earn at least that amount. The IRS stipulates that the funds allocated to the DCFSA must be used within the calendar year and the allowable grace period of the first 2.5 months of the following calendar year, before unused funds are forfeited (IRS, 2021b, p. 3; IRS, 2013). Unlike an HCFSA, which is subject to the uniform coverage rule that stipulates account holders can access all their allocated funds even before they are fully contributed, DCFSA draw only from available contributed funds for reimbursements. In practice, DCFSA account holders cannot use DCFSA funds before they have contributed those funds from their income and can only use the funds once the claim has been substantiated (IRS, 2022c).

In addition to the DCFSA, several other programs administered through the tax code relate to child and dependent care expenses. Both refundable and nonrefundable tax credits can help offset the taxes that a taxpayer owes. A nonrefundable tax credit enables the taxpayer to reduce the filer's tax liability for value of the credit, up to the amount owed. If the credit amount exceeds the amount that the filer owes, the taxpayer cannot access those excess funds. A refundable tax credit also enables a taxpayer to reduce a tax liability for the value of the credit; and if the credit amount is higher than the amount the taxpayer owes, the filer receives the rest as a refund (Tax Policy Center, 2020). The Child and Dependent Care Tax Credit (CDCTC) program is a nonrefundable tax credit that reduces a taxpayer's income tax liability. Like a DCFSA, the taxpayer can claim work-related care expenses, meaning the care is intended to help the taxpayer work or look for work. Eligible expenses also include in-home and out-of-home care expenses but exclude such educational expenses as tuition, camps, tutoring for children, or live-in nursing homes for dependents (IRS, 2021c).

Under the CDCTC taxpayers can claim up to $3,000 for one qualifying child under the age of 13 or dependent, or $6,000 for two or more qualifying children or dependents annually.[4] The amount by which the taxpayer can reduce a tax liability is then calculated by multiplying the qualifying expenses by a credit rate, which is based on the taxpayer's income. The credit

[3] Contribution limits included here are for tax year 2022. The American Rescue Plan increased these contribution limits to $10,500 for single taxpayers and married taxpayers filing jointly, and up to $5,250 for married taxpayers filing separately for tax year 2021.

[4] CDCTC amounts included are for tax year 2020 and again in 2022. For tax year 2021, The American Rescue Plan made the credit refundable and increased the maximum expenses to $8,000 for one qualifying child or dependent and $16,000 for two or more qualifying children/dependents. The American Rescue Plan also shifted the credit rate, raising it for filers with lower income, and reducing it for filers with higher income.

rate for taxpayers with a positive AGI less than $15,000 is 35 percent. The credit rate declines one percentage point for every $2,000 additional income, down to a credit rate of 20 percent for taxpayers with an AGI of $43,000 and higher (IRS, 2021c). There is no income limit to claim the credit (Crandall-Hollick, 2021).[5] The credit rate structure is shown in Figure 2.1. Eligible expenses are multiplied by the credit rate to determine the tax credit amount. For example, a taxpayer with an AGI of $40,000 with $3,000 of eligible expenses for one qualifying dependent could receive a credit of $660 on their taxes ($3,000 multiplied by 0.22, the credit rate for AGI $40,000). This credit reduces a filer's tax liability by offsetting taxes owed with the credit. Like other Americans, active-duty members and their families can apply eligible child and dependent care expenses to receive this credit. As we will discuss in Chapter Three, any contributions made to a DCFSA should a DCFSA be made available to service members, would offset or reduce the expenses that are eligible for the CDCTC.

FIGURE 2.1
CDCTC Credit Rates, Tax Year 2020

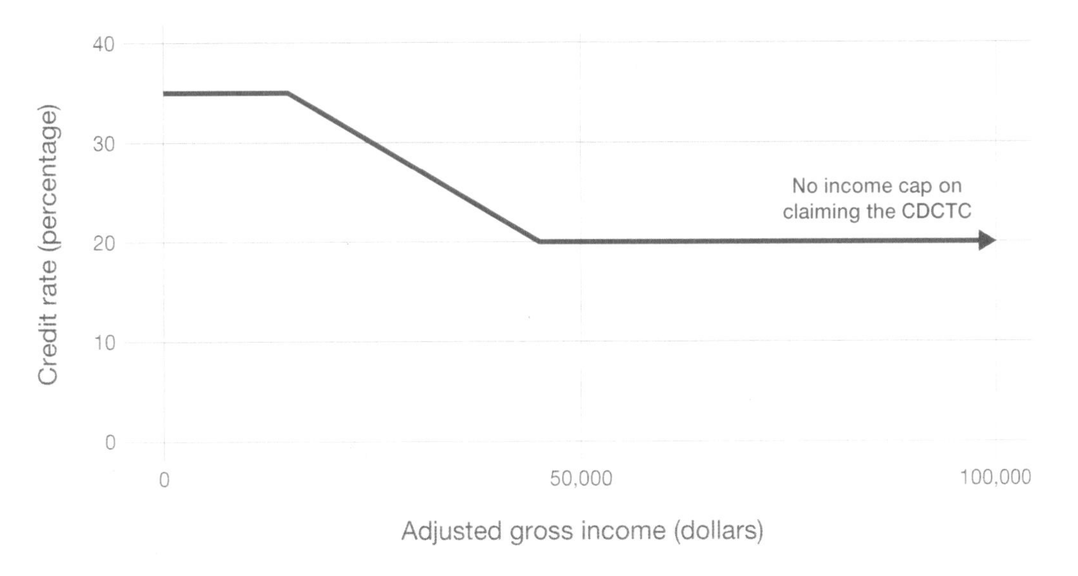

SOURCE: Authors' analysis of data drawn from Crandall-Hollick (2021).

---

[5] The American Rescue Plan changed the CDCTC for tax year 2021. It made the credit refundable, increased the maximum expenses to $8,000 for one qualifying child or dependent and $16,000 for two or more qualifying children or dependents. It also changed the credit rate. The credit rate in 2021 was 50 percent for filers with a positive AGI less than $125,000, and declined by one percentage point for every $2,000 until an AGI of $183,000, at which point the credit rate was 20 percent. Filers with an AGI $183,000 to $400,000, received a credit rate of 20 percent. The credit then phased out, declining by one percentage point for every $2,000 of income until the credit was $0 for filers with an AGI of at least $438,000.

The Child Tax Credit also reduces a family's tax liability and has the objective of providing support to families with children. Figure 2.2 shows the structure of the tax credit for married couples in tax year 2020. For every qualifying dependent child under the age of 17, a family receives a tax credit of up to $2,000, which phases out at $200,000 for unmarried filers and $400,000 for married couples. The tax credit is partially refundable for families whose tax liability is less than the value of the maximum amount, up to $1,400.[6] The Child Tax Credit is available to all taxpayers with eligible children, subject to the income restrictions to access the full amount of the credit (Crandall-Hollick, 2018). The Congressional Research Service estimates that 84 percent of all families with children receive some support from the Child Tax Credit (Crandall-Hollick, 2021). Active-duty members with children currently can access this credit through their tax returns.

Families with dependents who are not children can access the Other Dependent Tax Credit. This provides taxpayers with a $500 nonrefundable tax credit for each qualifying dependent, including children older than 17, or dependents living with the taxpayer, such as elderly parents (IRS, 2022g). Like the CTC, the Other Dependent Tax Credit phases out for high income earners above a set threshold. These credits are intended to provide support for families who are caring for dependents and can currently be claimed by active-duty service members.

Military members can also take advantage of the Earned Income Tax Credit (EITC). The EITC is a refundable tax credit that has the aim of supporting low- and moderate-income

FIGURE 2.2

Child Tax Credit, Married Couples, Tax Year 2020

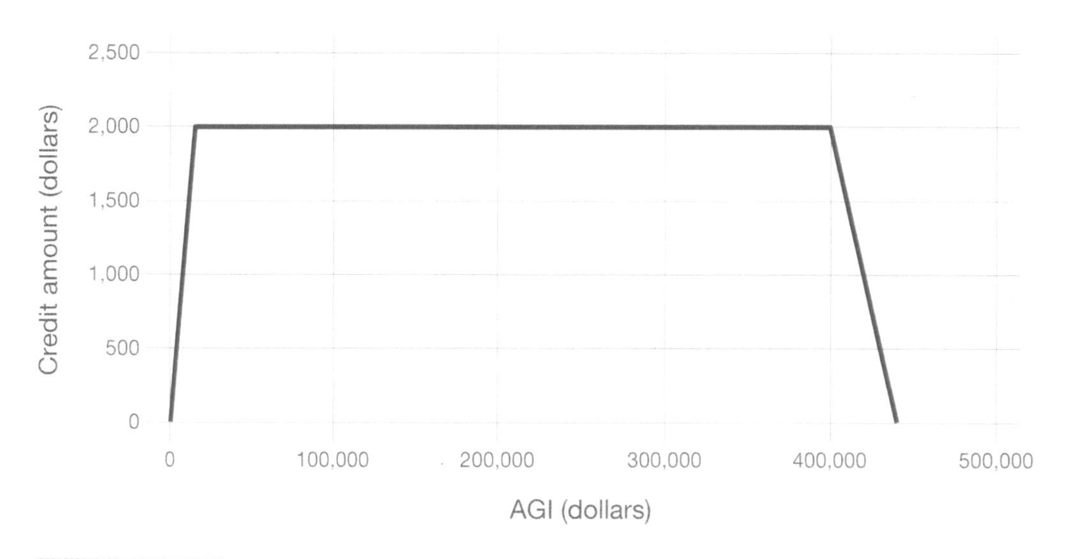

6 The American Rescue Plan expanded the CTC to $3,600 for children 0 to 5 years old, and $3,000 for children 6 to 17, and qualifying 17-year-olds for the first time. The amount phased down to the prior amounts at a particular income level ($150,000 for married filers and $112,500 for single filers), and it phases out for higher earners. The tax credit was made fully refundable, and it was administered monthly rather than annually.

working families and with investment income below a set threshold (IRS, 2022d; Garren, 2022).[7] The amount of the credit is based on a family's earned income and the number of children under the age of 19 (or age 24 if the young adult is a full-time student). The structure of the credit is shown in Figure 2.3 for unmarried filers in tax year 2020. The credit phases-in from the first dollar earned, plateaus until income exceeds the phase-out threshold; phase-in and phase-out rates and thresholds depend on the number of qualifying children, and whether filers are married or unmarried. For 2020, the maximum EITC for a taxpayer with one child was $3,584 per year; for two children, $5,920 per year; and for three or more children, $6,660 per year. A much smaller EITC is available to low-income workers without children. The structure of the tax credit mirrors the EITC for workers with children, but the maximum credit amount for a worker without children was $538 in 2020 (Crandall-Hollick, Falk, and Boyle, 2021).[8] Active-duty military members can claim the EITC and have particular circumstances that influence their credit amount. Military members can decide whether to include nontaxable military pay for the purposes of claiming the EITC. Military members can choose to include all or none of their nontaxable pay; married couples who both have

FIGURE 2.3
EITC Phase In and Phase Out, Unmarried Filers Tax Year 2020

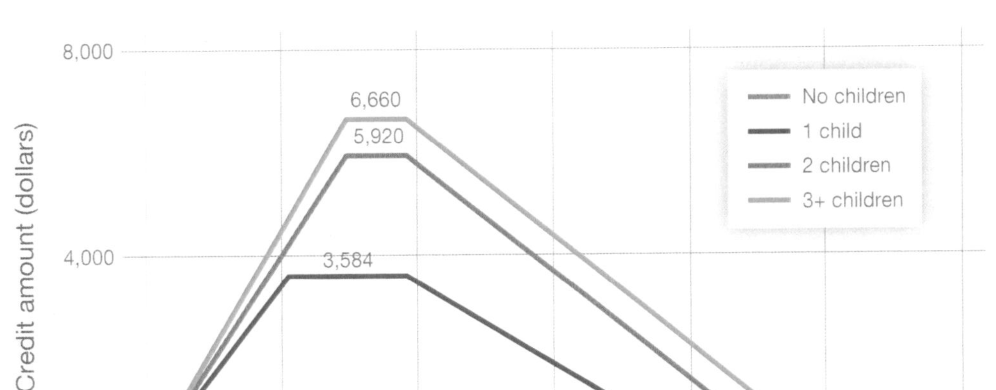

SOURCE: Authors' analysis of data drawn from Tax Policy Center (2021).

[7] The EITC is primarily targeted at low-income earners without significant assets. To claim the EITC, a taxpayer cannot have more than a certain amount of investment income which includes income from interest, dividends, capital gains, royalties, rental income or passive income. Prior to 2021, the limit was $3,650. The American Rescue plan increased this limit to $10,000 and will be indexed each year for inflation.

[8] This *childless* EITC was temporarily expanded under the 2021 American Rescue Plan, increasing the rate of the phase-in credit and also increasing the maximum credit amount to $1,502.

nontaxable military pay can include neither, one, or both individuals' noncombat pay (IRS, 2022d). Effectively, military members can calculate their tax credit under both scenarios and select the strategy that provides them with the highest credit.

Military Child and Dependent Care Benefits for Active-Duty Service Members and Families

DoD provides the largest employer-sponsored child care program in the country, providing and supporting care for 200,000 children of service members and civilian employees annually with an annual cost of over $1 billion (Kamarck, 2020). This program offers support on and off military installations for children ranging from infancy through 12 years old. The program involves four main services:

- child development centers (CDCs) or "on-base" care, mostly for children under age 6
- school-aged care (SAC) provides programs in stand-alone facilities, installation-based youth centers, or on-base CDC facilities for children ages 6 to 12 (or in kindergarten)
- family child care (FCC) that can be on or off base and is provided to children through age 12
- supplemental child care, which augments the other programs (the main program is a fee-assistance program that gives a child care subsidy for community-based [off-base] care that is paid by the services).

We briefly review each of these services in this subsection.

CDCs offer full-day and part-day child care for infants through preschool-aged children on DoD installations in facilities that are staffed and overseen by DoD—these are sometimes called installation-based care. The CDCs offer care during standard working hours, with some locations offering extended day schedules. CDC spaces are available to children of military families and DoD civilians alike, although spots are limited, and priority is given to military families. Military families where both spouses are working are also given priority over military families where one spouse is not working. Fees for CDCs are subsidized and are based on a family's total family income (TFI), which is composed of wages, salaries and tips from both spouses if married, and allowances, including the Basic Allowance for Housing and the Basic Allowance for Subsistence (Office of the Assistant Secretary of Defense for Manpower and Reserve Affairs, 2021).[9] These rates are shown in Table 2.4 for the 2021–2022 school year and range from $58 per week for families in the lowest income category, to $167

[9] TFI should be distinguished from taxable income. CDC fees are based on TFI as the text indicates, while the tax changes that we compute in Chapter Four are based on taxable income. An important way in which TFI differs from taxable income is the former includes allowances, while the latter does not; allowances are not subject to income tax. A further distinction is that neither TFI nor taxable income includes the tax advantage associated with receipt of allowances tax-free. Metrics of military pay, such as regular military compensation, typically recognize the tax advantage represents income to the member.

per week for military families earning over $140,000 (Office of the Assistant Secretary of Defense, Manpower and Reserve Affairs, 2021). As we discuss in Chapter Three, should a DCFSA be offered to service members, CDC fees paid by military families would be considered eligible expenses.

SAC programs are provided in installation-based youth centers or in CDC facilities. The programs are certified, and provide before and after school care, care during nonschool days

TABLE 2.4

Child Care Fees for CDCs and Community-Based Fee-Assistance Programs by TFI Category, School Year 2021–2022

Category	TFI	Weekly Fee Per Child
I	$1–$30,000	$58
II	$30,001–$40,000	$67
III	$40,001–$50,000	$82
IV	$50,001–$60,000	$100
V	$60,001–$70,000	$119
VI	$70,001–$80,000	$126
VII	$80,001–$90,000	$139
VIII	$90,001–$100,000	$142
IX	$100,001–$110,000	$146
X	$110,001–$120,000	$149
XI	$120,001–$130,000	$156
XII	$130,001–$140,000	$162
XIII	$140,001+	$167
DoD contractors and specified space-available patrons	Not applicable	$217
Standard hourly care	Not applicable	$7 per hour

SOURCE: Office of the Assistant Secretary of Defense, Manpower and Reserve Affairs (2021).

and during the summer. Rates are subsidized, and the cost to a family is determined by the family's TFI.

To expand the availability of child care options, given the wait lists for on-base care, and to increase accessibility for families who might prefer off-base care, the armed forces provide fee-assistance or a subsidy to service members to help cover off-base child care expenses if their children are enrolled in an approved community provider. To participate in the program, providers must be state-licensed, have had a state-licensing inspection within the previous year, ensure that all employees have passed FBI background checks, and be nationally accredited (Child Care Aware of America, undated-a).[10] Under the fee-assistance program, a family pays the approved community-provider the on-base CDC weekly rate associated with their family's TFI. The service's subsidy is the difference between the CDC rate and the rate charged by the provider, up to a cap established by each Service (Child Care Aware of America, 2021).[11] The caps for 2022 are:

- full-time care: $1,500 per month per child
- part-time care: $750 per month per child.

If the provider's fee is more than the established maximum allowable cap, a family is also responsible for covering the difference. Additionally, to be eligible for the full subsidy, both parents must be working or enrolled in school (Child Care Aware of America, undated-a).

The amounts that military families receive from the subsidy for community-based care would offset or reduce the maximum amount that can be contributed to a DCFSA or that is eligible for the CDCTC described above. As we will discuss further in Chapter Three, a military family receiving a community-based care subsidy at the monthly cap of, say, $1,500 per month for Army personnel, would reach the $5,000 per year cap on DCFSA contributions within a few months. Consequently, the maximum a member could contribute to the DCFSA would be zero.

Military families can also access FCC, also known as home day care, and they care for infants and school-aged children in their homes (either on or off an installation). They often offer a more flexible schedule to families, including full-day, part-day and extended care. These providers are limited to caring for, at most, six children at any given time (Kamarck, 2020). FCC providers are certified by installations or military services based on DoD requirements. The certification requirements differ slightly for providers based on and off military installations. For on-installation FCC providers to be certified, a provider must be state licensed, pass an annual inspection, and have either a Child Development Associate credential, an associate's degree or higher in early childhood education or child development, or accreditation by the National Association for Family Child Care (Child Care Aware of America, undated-a). Many FCC providers are military spouses (Kamarck, 2020, p. 26). Although FCC fee structures are determined by the providers themselves, military families who enroll

[10] There are locations where nationally accredited care is not available. In 2019, the military services launched a pilot where, in locations without nationally accredited care, providers could instead be approved if they are rated quality by their state's Quality Rating and Improvement System (QRIS). The program, Military Child Care in Your Neighborhood-Plus, is available in Maryland, Nevada, Virginia, and Washington as of 2022 (Child Care Aware of America, 2022a).

[11] Prior to 2021–2022, all services other than Army had a provider cap of $1,100 per month per child, $1,300 per month per child in high cost locations. Army had a provider cap of $1,500.

in FCC are eligible for fee assistance based on their TFI, such that they would pay the same weekly fee they would be charged for child care through an on-installation CDC or SAC.

DoD recently launched an in-home child care fee-assistance pilot. The pilot provides fee assistance to families who employ full-time, in-home child care, such as nannies. Providers must meet certain age, citizenship, and education requirements; have completed 32 hours of child care training; and have passed state and national background checks. The assistance can support care from 30 to 60 hours per week. Families are eligible for the assistance if they are active duty; in married households, families are eligible if they are dual active duty, or a spouse works or attends school full time. The program is currently offered in the District of Columbia area; Hawaii, San Diego, California; Norfolk, Virginia; and San Antonio, Texas. The payment structure is the same as the community-based fee-assistance program, and it is based on TFI (Child Care Aware of America, undated-a).[12]

Finally, although we focus here on military-provided dependent care programs, military families can also access community-based off-base care without a receiving a subsidy. Families might seek off-base dependent care and without fee assistance when an insufficient number of community-based care facilities have been approved to be a certified provider and on-base options are not available.

Summary

Active-duty service members and their families have access to health care and child and dependent care benefits through both DoD-sponsored programs and tax incentive programs. DoD provides programs in health care, including TRICARE and TRICARE Dental programs, that nearly eliminate out-of-pocket costs relating to accessing health care for members and keep costs low for family members. On-base DoD-subsidized child care provides a below-market rate option for care for working families. Members can access several tax credits to offset expenses related to child care and may be eligible for means-tested credits, such as the EITC, Child Tax Credit, or CDCTC that can reduce their overall tax burden. Together, these DoD-provided programs and tax credits may reduce eligible expenses for FSAs, or provide other methods by which active-duty families can address costs associated with health and child care.

[12] DoD also offers child care support for specific communities within the military. One such example is Respite Care. Through Respite Care, families of deployed service members, wounded warriors, and survivors of fallen warriors are eligible for up to 16 hours of child care per child per month, at no cost to the family. Army recruiters and ROTC Cadet Cadre are eligible for five hours of child care per child per month during their assignment. Another example is care provided for children enrolled in the Navy's Exceptional Family Member Program. These children are eligible for 40 hours of free care per month, provided in the child's home, in an FCC, or an off-base center. These programs are targeted at alleviating the particular care constraints that these families face.

CHAPTER THREE

How Might FSAs Financially Benefit Active-Duty Members and Their Families?

Important considerations for Congress and DoD in deciding whether to implement FSAs for service members is whether they impart a benefit to members and the cost and benefits to DoD of providing those benefits. In this chapter, we consider the context and mechanisms that influence how FSAs might benefit service members financially. This context underlies the computations of financial benefit provided in Chapter Four, and the analysis of costs and benefits to DoD explored in Chapter Five. Whether FSAs could impart a financial benefit depends on several factors:

1. **Extent to which active-duty members use health care and dependent care that is eligible for reimbursement through an FSA.** Information is needed on the extent of health care and/or dependent care expenses that would be eligible for an FSA, both the average amount of expenses per member and the share of members who would have these expenses. Related to the extent is the predictability of these expenses.
2. **How FSA options would affect other tax benefits and financial outcomes.** Information is needed on how expenses eligible for an FSA would interact with other tax-reducing programs and tax credits as well as other military programs (i.e., the net benefit taking account of these other programs).

How FSAs Could Financially Benefit Members and Their Families

FSAs provide a benefit to taxpayers in two ways. First, FSA contributions are excluded from the taxpayer's AGI on which taxes are assessed. Given that U.S. tax system is progressive, meaning that that the marginal tax rate rises with income, excluding income from the "top" of the taxpayer's earnings means that less income is subject to the top tax rate. For example, in tax year 2022 an unmarried taxpayer with an AGI of $45,000 would pay 10 percent on the first $10,275, 12 percent on income up to $41,775, and 22 percent on income over that amount up to $89,000 for a total tax liability of $5,517 (or $1,027.50 on the first $10,275 + $3,780 on income up to $41,775 + $709.50 on the last $3,225) (IRS, 2021b). If the taxpayer contributed $1,000 to an FSA,

AGI would be $44,000, and the amount taxed at the highest rate (22 percent) would be reduced. The taxpayer would owe $5,297 ($1,027.50 + $3,780 + $489.50) for a savings of $220.

Second, contributions to an FSA reduce the amount of earnings subject to Social Security and Medicare taxes, thereby reducing the payroll tax that the member pays.[1] The majority of service members have earnings below the Social Security wage cap, enabling them to save both Social Security taxes (charged at 6.2 percent of earnings) and Medicare taxes (charged at 1.45 percent of earnings) for a total of 7.65 percent of their FSA contribution. The payroll tax savings amounts to $382.50 if a service member contributes the maximum of $5,000 to a DCFSA and $218.03 if a service member contributes the maximum of $2,850 to an HCFSA. For service members with net earnings of FSA contributions greater than the Social Security wage cap ($137,700 in 2020), the service member would only save the Medicare portion of payroll taxes or 1.45 percent of the FSA contribution because he/she would pay the maximum amount of Social Security taxes irrespective of whether he/she contributes to an FSA.

This basic structure relies on a few assumptions. First, to be beneficial, the service member must incur eligible costs. Second, because of the use-it-or-lose it feature of FSAs, the predictability of these costs is also important, and, as discussed in Chapter One, the value of the benefit may be less than the dollar cost if members forfeit unused funds at the end of the plan year. Finally, the utility of the FSA depends on family earnings, access to tax credits, and any offsets that limit the contribution and benefit amount.

Health Care and Dependent Care Usage and Out-of-Pocket Expenses Among Active-Duty Members and Their Families

Health Care

Whether members and their families choose TRICARE Prime or TRICARE Select affects their health care expenses that would be eligible for an FSA. All active-duty service members were enrolled in TRICARE Prime, while 82 percent of active-duty families were enrolled in TRICARE Prime in fiscal year (FY) 2020 (Figure 3.1). Most of the remaining active-duty families, 17 percent, were enrolled in a TRICARE Select plan. The remaining were either nonenrolled or eligible for Medicare.

Table 2.2 in the previous chapter showed that average out-of-pocket costs are $0 under TRICARE Prime for active-duty members, while Figure 3.1 shows that all active members are covered by TRICARE Prime. Consequently, average expenses for accessing health care are virtually zero for service members. Family members enrolled in Prime plans are responsible

[1] Because the employer, in this case DoD, also pays Social Security and Medicare taxes for each service member, DoD would also benefit in the form of lower Social Security and Medicare contributions. We discuss this savings in Chapter Five. Reducing a service member's earnings subject to Social Security taxes would also potentially reduce his or her earnings used to calculate future Social Security benefits, which are based on an individual's 35 highest years of earnings.

FIGURE 3.1
Health Plan Coverage of Active-Duty Members and Their Families, 2020

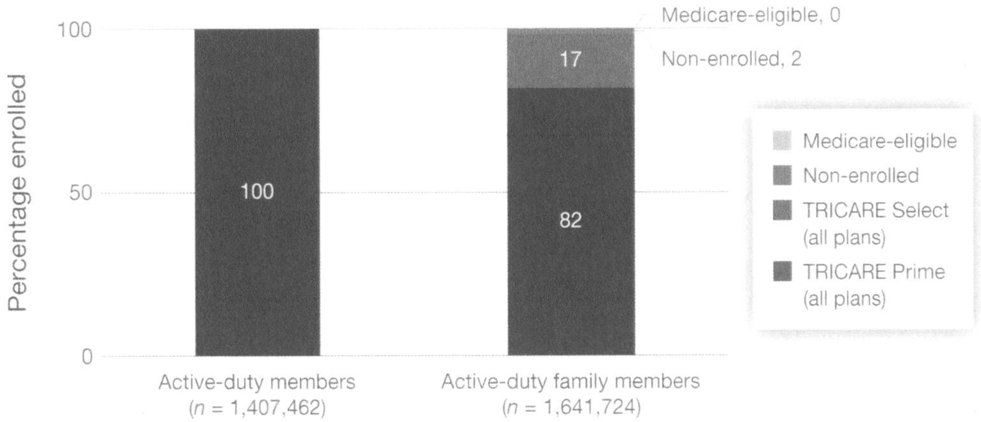

Percentage enrolled
100
50
0
Medicare-eligible, 0
Non-enrolled, 2
17
100
82
Medicare-eligible
Non-enrolled
TRICARE Select (all plans)
TRICARE Prime (all plans)
Active-duty members (*n* = 1,407,462)
Active-duty family members (*n* = 1,641,724)

SOURCE: Authors' analysis of data drawn from Office of the Assistant Secretary of Defense for Health Affairs, Defense Health Agency (2021, p. 35).
NOTE: Numbers do not sum to 100 due to rounding.

for co-payments only when they seek care without a referral, potentially facing higher out-of-pocket costs than service members depending on personal choices for sought-after care. For active-duty family members enrolled in Prime, the annual out-of-pocket cost hovered around $100 per year for 2018 through 2020 (Figure 3.2), below the comparable out-of-pocket costs faced by civilians enrolled in HMO plans (Office of the Assistant Secretary of Defense for Health Affairs, Defense Health Agency, 2021). Thus, for the 82 percent of active-duty families who are covered by TRICARE Prime, their average out-of-pocket health care expenses were about $100 per year. For the 17 percent of active family members enrolled in TRICARE Select, average out-of-pocket expenses to cover co-payments and cost-sharing were about $500. Although higher than the out-of-pocket expenses expected under TRICARE Prime, the annual cost of TRICARE Select is significantly lower than comparable costs faced by civilians in PPO plans (see Figure 3.3).

Other, nonhealth plan–related health care expenses might be eligible for FSA coverage. Specifically, many over-the-counter medications and supplies are eligible expenses. We have no data on such expenses by active military members or their families, but the Consumer Expenditure Survey that the Bureau of Labor Statistics administers and collected by the Census Bureau provides this information for U.S. households in general. Americans, on average, spent $387 in 2020 on over-the-counter medications and supplies, including nonprescription drugs and vitamins, shown in Table 3.1. Although some of the medical supplies included in the survey might be covered by TRICARE health plans for military personnel, other expenses might be less relevant for the military community (adult diapers), a population that is younger and healthier than the country as a whole. Other expenses could be more relevant to military personnel, given the physically taxing aspects of military service.

FIGURE 3.2
Out-of-Pocket Costs by Type for Active-Duty Families Enrolled in TRICARE Prime Compared with Civilian HMO Costs, FYs 2018–2020

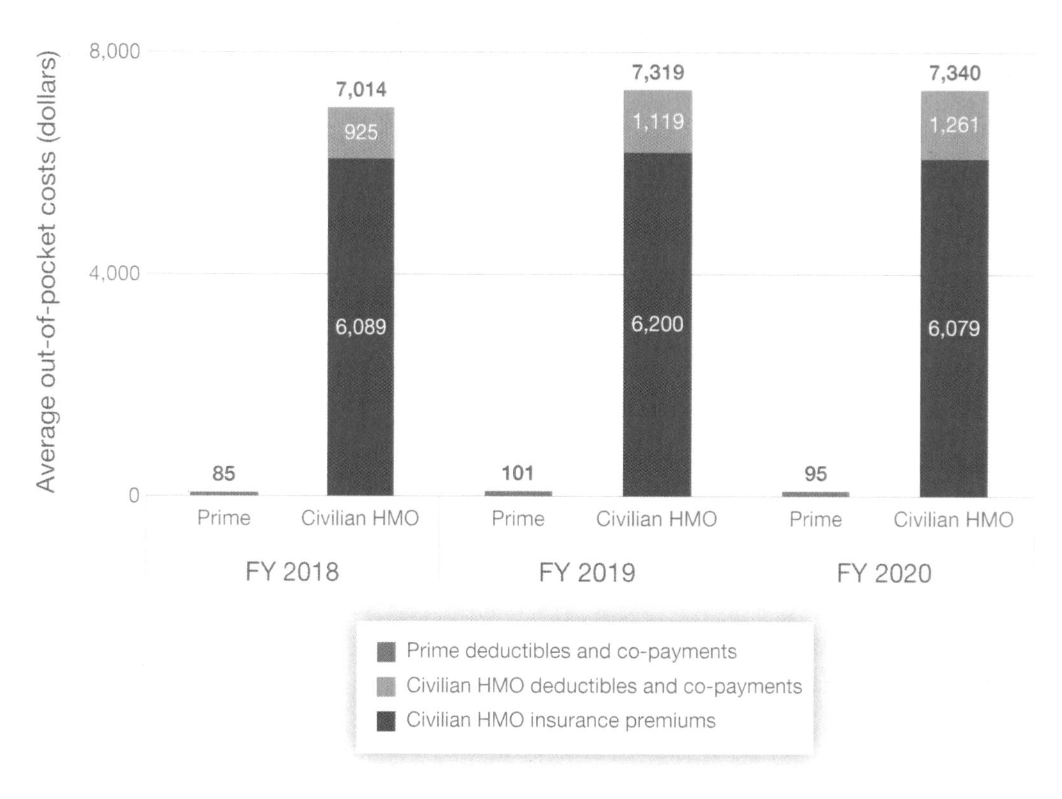

SOURCE: Authors' analysis of data drawn from Office of the Assistant Secretary of Defense for Health Affairs, Defense Health Agency, (2021, p. 215).

Although it is difficult to surmise to what degree the average expenses for the general population are relevant for military personnel, these estimates provide a helpful benchmark for understanding how much military families might be spending on health-related expenses outside their health plan.

Eligibility of Out-of-Pocket Health Care Costs

Ultimately, to financially benefit from an HCFSA, service members would need to incur out-of-pocket costs that are eligible for FSA coverage. Table 3.2 provides a summary of the types of expenses incurred by service members and their families that are eligible for FSA funds. The table suggests that dental care fees could be a source of out-of-pocket costs for military families and some of the costs could be large. For example, the average cost of orthodontia is large (about $3,100 according to Hung et al., 2021).

FIGURE 3.3
Out-of-Pocket Costs by Type for Active-Duty Families Enrolled in TRICARE Select Compared with Civilian PPO Costs, FYs 2018–2020

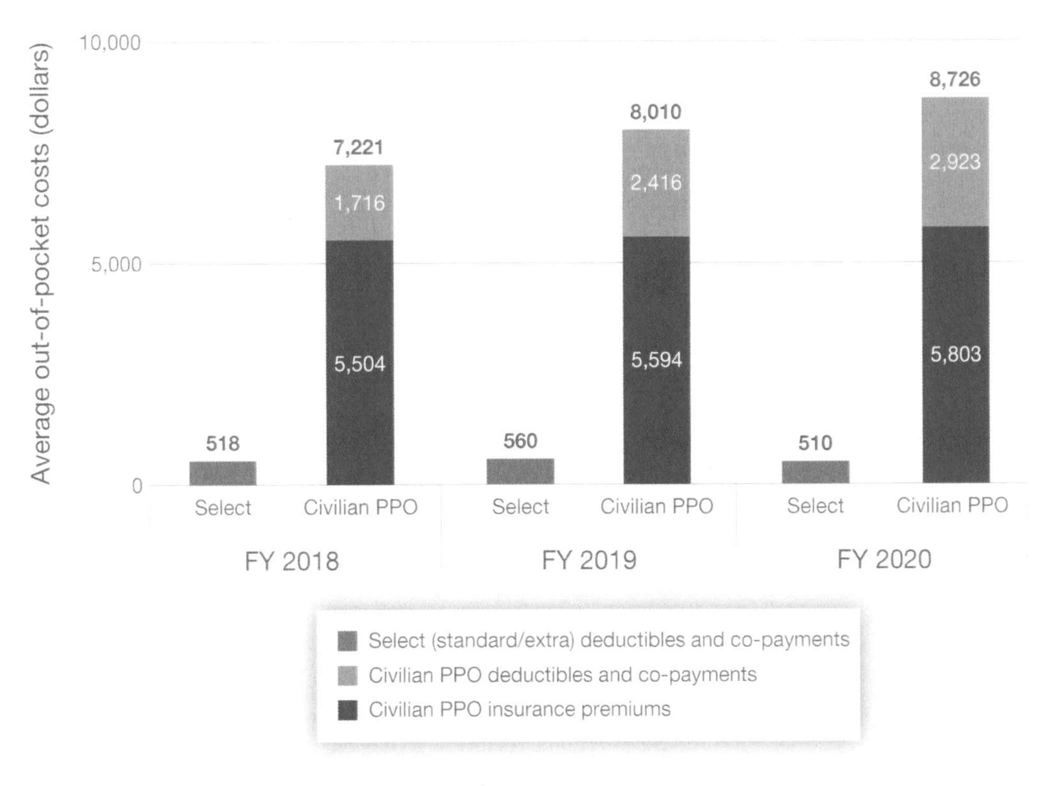

SOURCE: Office of the Assistant Secretary of Defense for Health Affairs, Defense Health Agency (2021, p. 217).
NOTE: Numbers do not sum to the totals due to rounding.

Child Care

Whether families participate in DoD-sponsored programs determines the child care expenses they incur and the eligibility of those expenses for FSA reimbursement. Determining the number of families participating in DoD-supported programs is challenging because data are reported on children enrolled in a program but not on members with enrolled children. Table 3.3 presents two measures of enrollment in DoD programs. The first column presents the number of children enrolled in these programs on a date of record in September 2019.[2] The first column likely underestimates the number of children who participate in these programs throughout the tax year, because the data capture a point-in-time measure of enroll-

2 Table 3.3 presents 2019 enrollment data provided to the research team by the Office of Military Community and Family Policy. We also reviewed 2018 and 2020 enrollment data. Levels for 2018 enrollment were similar to 2019 levels. Enrollment during 2020 is notably lower, likely because of the coronavirus disease 2019, during which safety requirements limited availability and families' routines and demands for child care were disrupted.

TABLE 3.1
Average Annual Expenses from the Consumer Expenditure Survey, 2020

Expense Category	Average Annual Expense
Nonprescription drugs	$115.74
Nonprescription vitamins	$101.46
Medical supplies	$169.70
Eyeglasses and contact lenses	$74.12
Hearing aids	$27.52
Topicals and dressings	$48.42
Adult diapers	$2.22
Medical equipment for general use	$9.10
Supportive and convalescent medical equipment	$6.13
Rental of medical equipment	$1.15
Rental of supportive, convalescent medical equipment	$1.05
Total: nonprescription drugs, vitamins and medical supplies	$386.90

SOURCE: Bureau of Labor Statistics (2020).

NOTE: Data include all consumer units, including households, financially independent individuals living with others, and two or more people cohabitating and making joint financial decisions.

ment as families access care arrangements at different times through the year. The second column shows the number of children enrolled in programs throughout FY 2019, but the number may overestimate participation if it captures children who have changed child care settings because of a permanent change of station or another event. Additionally, we present the number of children who received fee assistance for a community-based care provider. These data suggest that 9 percent to 16 percent of children enrolled in DoD-supported child care are enrolled in community-based fee-assistance programs.

For service members who participate in either installation-based care (e.g., CDC, SAC, or, in some cases, FCC) or who access community-based fee-assistance care, child care costs are based on TFI, as described in Chapter Two. Figures 3.4 and 3.5 show the distribution of enrolled children in a CDC or SAC, respectively, by TFI category during the 2021–2022 school year.[3] (Table 2.4 defines income categories. As noted, TFI consists of wages, salaries, tips and allowances, including the Basic Allowance for Housing and the Basic Allowance for Subsistence.) CDC enrollment is highest among those in Categories VIII, IX, and XIII, corresponding to incomes between $90,001 and $100,000, $100,001 and $110,000, and more than $140,000, respectively. Across the categories, 17 percent of CDC-enrolled children are

[3] The tabulations shown in these Figures are based on the date of record data for November 2021.

TABLE 3.2

Health Care Costs for Active-Duty Service Members and Families by Whether Expenses are FSA-Eligible

FSA Eligibility of Expenses	Health Care Costs
Incurred health care costs ineligible for FSA funds	Medical care: • Enrollment fee (relevant for retirees and their families; active military members and their family members are exempt from an enrollment fee for TRICARE Prime and TRICARE Select) Dental care: • TRICARE Dental Plan monthly fees (relevant only for enrolled family members)
Incurred health care costs eligible for FSA funds	Medical care (all plans unless otherwise indicated): • Point-of-service fees for care accessed outside a referral (TRICARE Prime only; relevant only for enrolled family members) • Co-payments and cost shares for health care (TRICARE Select only) • Per diem for inpatient care (relevant only for enrolled family members) • Prescription drugs accessed outside a military pharmacy • Over-the-counter medicines Dental care: • Cost shares (relevant only for enrolled family members)

SOURCE: IRS, 2022b.

TABLE. 3.3

Estimates of Enrollment in DoD-Sponsored Child Care Programs

Child Care Program	Date of Record Enrollment (September 2019)	Annual Enrollment (FY 2019)[a]
CDC	60,960	116,538
FCC	4,714	8,381
SAC	29,110	51,294
Community-based fee assistance	36,074[a]	36,074
Total	130,858	212,287

SOURCES: Office of Military Community & Family Policy, undated; U.S. Government Accountability Office, 2021c.

NOTES: Columns reflect different methods for counting enrollment. *Date of Record Enrollment* reflects the number of children in a program on a given day, which may underestimate the total number of children enrolled in programs throughout the year. *Annual Enrollment* reflects the number of children enrolled in programs throughout the year but may duplicate counts of children who enrolled in different programs or locations through the year.

[a] Reflects community-based fee assistance for full FY.

FIGURE 3.4

Number of Children Enrolled in a CDC by Age and TFI Category, November 2021

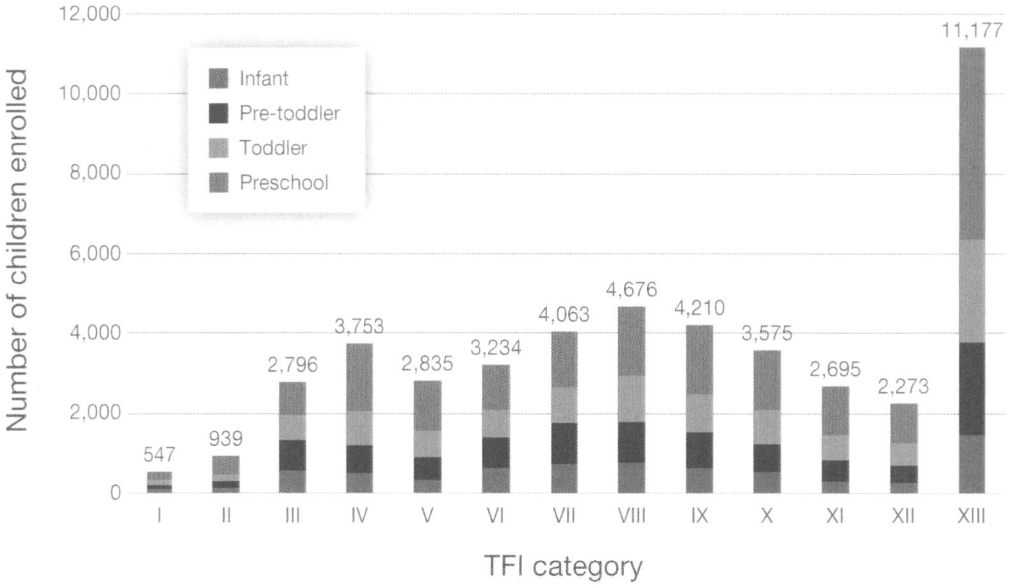

SOURCE: Author's analysis of TFI data provided to the research team by the Office of Military Community and Family Policy.

NOTE: TFI categories are defined in Table 2.4.

from families with total income of less than $60,000, 32 percent from families with between $60,000 and $100,000 in total income, 27 percent from families with between $100,001 and $140,000, and 24 percent from families with more than $140,000 in TFI. As discussed in Chapter Two, fees are set based on TFI. A family with $55,000 in TFI and an infant enrolled at a CDC would pay $100 weekly (TFI Category IV) or $5,200 annually in child care costs.

Unsubsidized child care expenses are higher on average than child care accessed through military-supported programs and can vary dramatically across locations. Child Care Aware of America estimates that the national average for a year of care for children under the age of 5 is around $10,174 (Child Care Aware of America, 2022a, p. 42). The average is higher for infants than for preschool-aged children and for center-based programs than in home-based child care. In 2020, the annual average cost of center-based care for an infant in California was $17,384, $12,168 for a 4-year-old, and $14,399 for before- and after-school care for school-aged children. In Texas, average annual costs were estimated to be $10,826 for infant care, $9,147 for a 4-year-old's care, and $5,949 for SAC. In Alabama, one of the lowest-cost states for child care, the average annual cost for infant care was estimated to be $7,800, and care costs for a 4-year-old and school-aged children were estimated at $7,280 annually (Child Care Aware of America, 2022b, pp. 2–3).

FIGURE 3.5

Number of Children Enrolled in SAC Programs by Age and TFI Category, November 2021

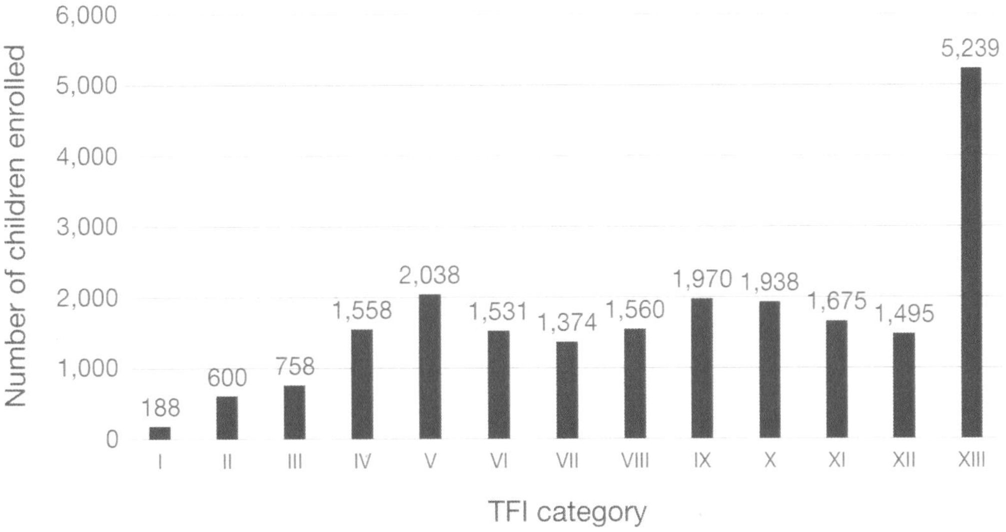

SOURCE: Author's analysis of TFI data provided to the research team by the Office of Military Community and Family Policy.
NOTE: TFI categories are defined in Table 2.4.

Eligibility of Out-Of-Pocket Dependent Care Expenses

How and whether service members and their families can financially benefit from a DCFSA depends not just on whether expenses are eligible for a DCFSA. It also depends on a family's earnings and tax credits, their employment status, and, critically, on the type of dependent care it receives, as we describe in this subsection. Ineligible and eligible expenses for child and dependent care FSAs are included in Table 3.4.

Employment and earnings matter because DCFSA funds can cover only work-related care expenses. This means that DCFSA funds can only cover care while the service member is working, and, if relevant, the member's spouse is working, looking for work, or enrolled as a student. Additionally, DCFSA contributions cannot exceed the lower-earning spouse's earned income. Thus, if a family wanted to max out its contributions to a DCFSA at $5,000, the lower-earning spouse's earning would need to exceed that amount. For active-duty married families to benefit from a DCFSA, the dependent care must support a family's work or studies, and a family must have earned income.

Importantly, the type of dependent care will affect the financial benefit of a DCFSA for military families. Families who enroll their children in on-base child care programs (such as a CDC or SAC) can use DCFSA dollars to cover their weekly fees. Although these programs are subsidized by the military, this subsidy is not considered a direct subsidy for active-duty

TABLE 3.4

Dependent Care Costs for Active-Duty Service Members and Families by Whether Expenses are Eligible for an FSA

FSA Expense Eligibility	Child and Dependent Care Costs
Incurred costs ineligible for FSA funds	Child care: • Child care for nonwork-related time • Educational fees (e.g., tuition for kindergarten or higher grades; tutoring) • Extracurricular activities • Overnight camps Dependent care: • Nursing home
Incurred costs eligible for FSA funds	Child care: • Care for a child or children under the age of 13 while parents are working, actively looking for work, or full-time students (including care provided through a formal program, or in-home care from a nanny or babysitter) • Day care and preschool • Before- and after-school care • Summer day camp Dependent care: • Care for a spouse or dependent (13 years or older) who is physically or mentally unable to care for themselves • Adult day care center • Daily elder care All care: • Payment processing fees • Transportation to and from care provided by the care provider

SOURCE: IRS, 2021a.

military members.[4] Figures 3.4 and 3.5 show that 91 percent of preschool-aged and young children in installation-based CDCs and 93 percent of children in SAC are from military families in TFI Category IV or higher. For these children, out-of-pocket child care expenses exceed $5,000, meaning that their parents could contribute the maximum of $5,000 to a DCFSA if they meet all other eligibility criteria. Similarly, families who access community-based child care without fee assistance can use a DCFSA to cover their weekly expenses up to the annual maximum of $5,000, assuming their child care expenses are eligible for a DCFSA.

In contrast, families who enroll their children in subsidized community-based care would unlikely be able to use a DCFSA because the fee assistance would offset eligible DCFSA expenses, according to the U.S. tax code (IRS, 2022g, p. 11). Although DoD pays the subsidy

[4] Children of DoD civilian sponsors who enroll in these programs are considered to be receiving a direct subsidy. Therefore, the value of the subsidy will offset DCFSA contributions for civilians but not for military personnel. Although we do not know Congress's intent in making this distinction between civilians and military personnel, it is possible Congress wanted to provide what appears to be an additional benefit to military members in recognition of their military service.

directly to the provider, the subsidy offsets the eligible expenses for a DCFSA dollar for dollar. Any combination of child care subsidy and DCFSA contributions cannot exceed the maximum DCFSA contribution limit of $5,000. The limit holds even if different care arrangements are used. To illustrate, if a family accesses at least $5,000 in fee assistance for care during the school year, a family is ineligible to contribute or use DCFSA funds for summer care, such as day camp. It is likely that families receiving fee assistance for community-based care will receive a subsidy that will offset any contribution to a DCFSA. Take for example a family with a TFI of $55,000 (TFI Category IV) and has child care expenses of $20,000 annually. If receiving fee assistance, a family would pay $100 weekly given that the family is in Category IV in Table 2.4 (or $5,200 annually assuming continuous enrollment), and the military, and specifically the member's service would pay the remainder of the fee directly to the provider. The services cap their contributions to providers and subsidize for care up to $1,500 monthly, or $18,000 annually, less the fees paid by families based on their TFI. In the scenario provided here, a family with $20,000 in annual child care expenses would have paid $5,200 for fees set according to its TFI, and the military would have paid $12,800 directly to the provider—up to the provider cap—an amount that far exceeds, and thus offsets, the contribution limit of $5,000 for a DCFSA. Furthermore, in this example, a family would have an additional $2,000 for the additional cost above the provider cap (equal to $20,000 – $5,200 – $12,800) of out-of-pocket expenses that could not be covered by the DCFSA.

Therefore, how and whether a military member would benefit from a DCFSA depends crucially on the type of dependent care and, more importantly, whether it is on-base care or subsidized community-based care. Members with children in community-based fee-assistance programs would be unlikely to benefit from a DCFSA, while those receiving on-base care, such as in a CDC, would benefit. Furthermore, because child care spaces in the CDCs are supply constrained (Kamarck , 2020), not every military family who would prefer to receive care on base (and fully benefit from the DCFSA) are able to do so. Furthermore, the DCFSA could exacerbate the shortage of on-base care spaces as more military families attempt to seek on-base care to take advantage of the DCFSA, an advantage that is less likely for off-base fee-assistance care. Although they may find certified off-base providers where expenses are eligible for fee-assistance, this option would most likely eliminate the advantage of having a DCFSA. Concerns about equity in access to the benefits of a DCFSA could potentially be addressed by eliminating the offset for the employer subsidy of child care. Were the offset to be eliminated, a DCFSA could be used to cover any remaining costs not covered by the DoD subsidy. Legislation would need to be enabled to make an employer child care subsidy and DCFSA stackable. We return to this point in our concluding chapter.

Interaction Between FSAs and Other Tax Benefits

As mentioned at the start of this chapter, whether FSAs could impart a financial benefit to service members depends on a second factor—specifically, how FSAs interact with other programs and specifically other tax benefits. FSAs enable taxpayers to exclude eligible health and dependent care expenses from income that is subject to taxation, as discussed earlier in the chapter. However, FSAs, exclusions of taxable income, and tax credit programs interact in important yet complicated ways that can affect the ultimate financial benefit of having an FSA.

Table 3.5 presents an overview of these interactions. Pre-tax contributions to FSAs are not included in earned income and are thus not included in a taxpayer's AGI. By making these contributions, taxpayers might find themselves eligible for different amounts of tax credits than they otherwise would have been. Reducing a taxpayer's AGI might change whether the taxpayer falls above or below credit thresholds, change the credit rate calculation, and determine the amount of the credit. The adjustments interact with other tax credits and so the benefit varies at different income levels.

Figure 3.6 illustrates the complicated interaction between an FSA and the EITC and the implications of FSA contributions for income taxes. Consider a taxpayer who contributes

TABLE 3.5

Interactions Between FSAs and Tax Benefits

Tax Benefit	How an FSA Interacts with or Affects the Tax Benefit
CDCTC	DCFSA: Individuals can use both the CDCTC and a DCFSA with two important constraints: • Expenses paid for by the FSA cannot also be claimed through the CDCTC. • For every dollar contributed to a DCFSA, the taxpayer must reduce the maximum amount of qualifying expenses for the CDCTC. Therefore, the taxpayer will be able to benefit from both programs, only if the federal tax credit is larger than what is contributed to the DCFSA. All FSAs: The maximum CDCTC amount is calculated based on the taxpayer's AGI. Contributions to an FSA will lower the taxpayer's AGI and affect the credit rate calculation.
Child Tax Credit and the Other Dependent Tax Credit	All FSAs: The Child Tax Credit and Other Dependent Tax Credit are calculated based on the taxpayer's AGI. Contributions to an FSA will lower the taxpayer's AGI and affect the amount of the tax credit.
EITC	All FSAs: The EITC calculated based on the taxpayer's taxable earnings. Contributions to an FSA will lower the taxpayer's earned income and affect the amount of the tax credit. Special considerations for service members: Service members can choose to include or exclude nontaxable pay for the purposes of the EITC. This decision would also be affected by FSA contributions that would reduce the service member's earned income.
Itemized health care deductions	HCFSA: Individuals cannot itemize deductions for health care expenses paid for by an FSA. All FSAs: Contributions to an FSA lower the taxpayer's AGI. Because individuals are able to deduct medical expenses that exceed 7.5 percent of the AGI, lowering their AGI through FSA contributions could mean that taxpayers could deduct more expenses than they would have been able to deduct had FSA contributions not been excluded. However, expenses that are paid through the FSA cannot be itemized, and this factor reduces the amount that is deductible.

SOURCES: IRS, 2021a; 2022b; 2022d; 2022f; 2022g.

FIGURE 3.6
Example of the Interaction Between an FSA and the EITC for Unmarried Member with One Child, Tax Year 2020

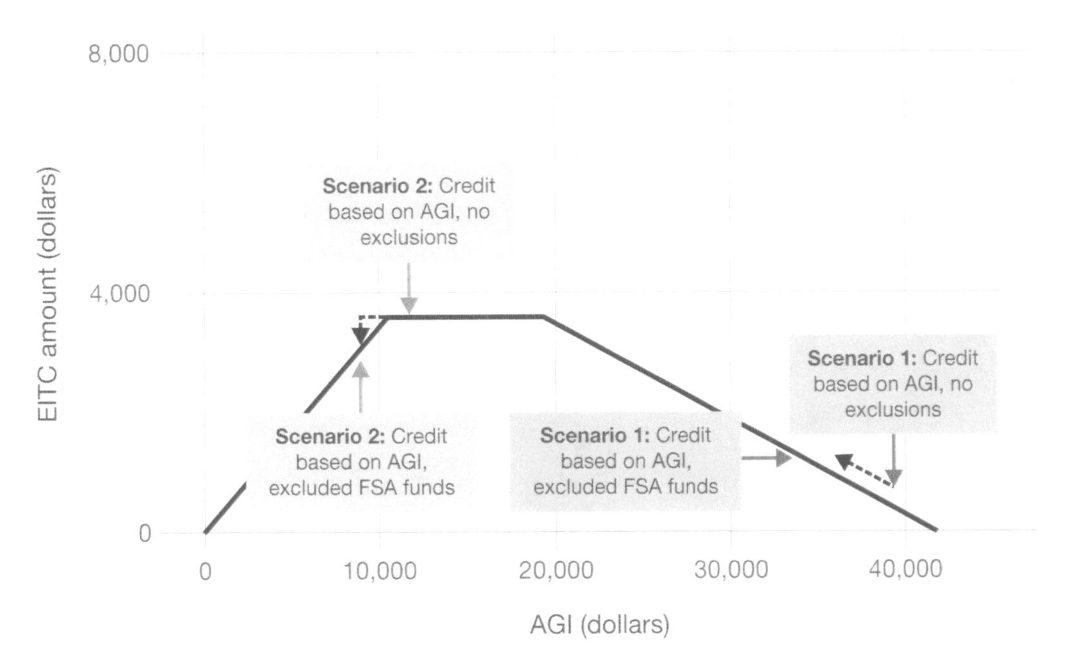

to an FSA and who excludes that income from earned income. In the absence of this exclusion, the taxpayer may have been at a particular part of the EITC's phase-out range or even just beyond the eligibility threshold to claim the EITC. With the exclusion and the implied lower earnings, the taxpayer may become eligible for a larger credit than the individual otherwise would have been. This case is demonstrated for AGI over $19,330 in Scenario One in Figure 3.6 when the taxpayer is in the "phase-out" range or beyond (Tax Policy Center, 2021). If instead, a person's pre-FSA earned income placed the individual in the phase-in region of the EITC (AGI less than $10,000 in Figure 3.6), then reducing earned income by the amount contributed to an FSA would reduce the amount of EITC that could be claimed. Finally, if a person's pre-FSA earnings places him or her on the plateau range of the EITC, then contributing to an FSA could leave the amount of EITC unchanged or even reduce it, shown in Scenario 2 in Figure 3.6. The effect of an FSA on the Child Tax Credit is similar to that described for the EITC. Reducing AGI by the FSA contribution amount can increase, decrease, or leave the amount of Child Tax Credit unchanged depending on whether post-FSA earnings move the service member along the phase-out range, plateau, or phase-in range of the credit. In contrast to the EITC and Child Tax Credit, contributions to a DCFSA will reduce the CDCTC that a service member can claim. This is because each dollar contributed to a DCFSA crowds out dollar for dollar the amount of eligible expenses that can be claimed for the CDCTC.

The overall net effect on tax liability from participating in an FSA program and, therefore, the net financial benefit of an FSA to service members is ambiguous and cannot be predicted a priori. The complex interactions between FSA contributions and certain credits mean that determination of the net benefit must be empirically estimated. In Chapter Four, we empirically estimate the change in taxes from participating in FSAs. After explaining our methodology for our empirical estimates, we also provide more detailed examples of how contributions to an FSA program interacts with other tax programs to affect the overall change in members' tax liability.

CHAPTER FOUR

Estimating Tax Benefits of DCFSAs and HCFSAs

As discussed in Chapter Three, the net effect of an FSA on total taxes is ambiguous due to its interaction with certain credits. Contributions to an FSA reduce earnings subject to income taxes and payroll taxes, thus yielding a tax savings from reducing the amount of income tax before credits (i.e., the amount of tax calculated based on taxable earnings) and reducing the amount of pay subject to Social Security and Medicare payroll taxes. However, certain tax credits—specifically, the EITC and the refundable portion of the CTC—are calculated based on taxable earnings. Because of the nonlinear schedules for these credits discussed in Chapter Two, reducing taxable earnings through contributions to an FSA might decrease, increase, or have no effect on the amount of EITC or Child Tax Credit that a service member could claim. Moreover, contributions to a DCFSA reduce the CDCTC that a service member could claim because DCFSA contributions offset dollar for dollar the amount of expenses eligible for the CDCTC.

We conduct simulation analysis to investigate the effect of FSAs on total taxes for active-duty service members under a variety of scenarios. We estimate the effect of FSAs on taxes separately for DCFSAs and HCFSAs and then present results from participating in both a DCFSA and an HCFSA. Before presenting the results, we first describe our methodology.

Methodology

To better understand the relationship between service member characteristics and tax implications of FSA contributions, we estimate how taxes would change under different scenarios using the National Bureau of Economic Research (NBER) TAXSIM (Feenberg and Coutts, 1993).[1] NBER TAXSIM is a collection of programs and data that implements a microsimulation model of U.S. federal and state income tax returns for each filing year. Unlike off-the-shelf software, such as TurboTax, NBER TAXSIM takes user inputs and calculates income tax liability and tax credits in batches. The batch calculation capability allows us to calculate the

[1] The NBER TAXSIM model and documentation can be online (NBER, undated).

tax change from contributing to an FSA for service members with different characteristics and income assumptions in a single batch.

For our analysis, we specify the following user inputs:

- tax year
- service member earnings subject to taxation
- marital status
- spouse earnings (if service member is assumed to be married)
- number of children by age group
- child care expenses applied to the CDCTC.

The scenarios we consider vary these attributes of the service member and family. All simulations are estimated assuming 2020 tax year law.[2] We set spouse earnings equal to $25,000 for married service members when we conduct simulations by member paygrade.[3] However, as we discuss later in this chapter, we also conduct the simulations by different levels of family income for us to consider scenarios in which spouse earnings can be higher or lower than $25,000. The simulations by family income also allow us to consider cases in which the service member's taxable earnings vary because of the receipt of special and incentive pay. The tax simulation calculations are presented separately by marital status and number of children (i.e., the term *number of children* refers to the number of children under age 13).

To estimate the effect of FSAs on taxes, we need to specify the amounts contributed to an FSA. For the DCFSA analysis, the bulk of the analysis will assume that service members contributed the maximum of $5,000 to a DCFSA. In Chapter Three, we showed that average child care expenses in the United States far exceed the $5,000 cap and that the cost of care for most children covered by installation-based care also exceeds the $5,000 cap, supporting the assumption that participants in a DCFSA would likely contribute the maximum amount allowed. For the HCFSA analysis, we examine the effects of contributing two different amounts, $500 and $2,850. Data presented in Chapter Three showed that average out-of-pocket medical expenses for active-duty families covered by TRICARE Select was about $500. We also present results assuming HCFSA expenses of $2,850, the maximum allowed in 2022, to demonstrate the potential benefit from contributing to an HCFSA when medical expenses are expected to be high due to anticipated expenses, such as orthodontia, dental, eyewear, and so forth.

NBER TAXSIM does not explicitly include FSAs in its calculations. As a result, when we estimate the effect of FSAs on taxes, we deduct the amount contributed to the FSA from

[2] We chose to use 2020 tax year because recent expansions of various tax provisions, including the expansions of the CDCTC and DCFSAs for the 2021 tax year, are temporary and scheduled to revert back to 2020 tax law unless extended.

[3] Using Social Security earnings records, Burke and Miller (2018) found that annual military spouse earnings among working military spouses was $22,800 in 2013. Inflating this amount to 2020 dollars yields $25,000.

service member earnings. We also assume that the service member would spend the entire amount contributed to an FSA. For DCFSAs, we also deduct the amount contributed to the DCFSA from the child care expense input variable, so the correct amount is used to calculate the CDCTC. To estimate the change in total service member taxes from contributing to an FSA, we manually add in the reduction in payroll taxes. For service members with earnings below the 2020 Social Security wage cap of $137,700, the service member (and DoD as discussed in Chapter Five) saves on both Social Security taxes (charged at 6.2 percent of earnings) and Medicare taxes (charged at 1.45 percent of earnings) for a total payroll tax savings of 7.65 percent multiplied by the FSA contribution. Service members with net earnings above the Social Security wage cap even after the FSA contribution is deducted from earnings do not experience a Social Security tax savings from contributing to an FSA because he/she pays the maximum Social Security taxes irrespective of contributing to an FSA. Service members with earnings just above the Social Security wage cap may experience Social Security tax savings on a portion of their FSA contribution if the earnings net of the contribution is below the Social Security wage cap. We account for each of these situations in our estimates of the payroll tax savings to the service member in the analysis here.

We use NBER TAXSIM to conduct simulations by paygrade and by gross earnings to examine how contributing to an FSA affects taxes from two different viewpoints. In both cases, we estimate the change in tax liability assuming that service members choose to participate in FSAs. In practice, service members who would experience a tax increase from contributing to an FSA would choose not to participate, causing FSAs to have no impact on their taxes.

The first viewpoint estimates the change in tax liability from contributing to an FSA for a typical service member in each paygrade to provide information on which service members might benefit from contributing to an FSA. We could also do these simulations by grade and year of service, in keeping with the structure of the military pay table, but to reduce the number of computations, we consider average earnings within each paygrade using information on the most commonly observed year of service within each grade. More specifically, to determine how to set service member earnings, we used counts of service members by paygrade and years of service from the September 2021 Active Duty Master File to determine which years of service bin had the greatest percentage of service members in each paygrade. For each paygrade, we then set service member earnings equal to the 2020 annual basic pay rate for the most-common observed years of service bin as shown in Table 4.1. In the analysis by paygrade, we first present specific examples to walk through how contributing to an FSA affects the different components of total tax and then show our estimates of total tax changes by paygrade. In the examples, a positive change can represent a tax increase or a tax decrease *depending on the tax component*. For example, a positive change in a tax credit means that there is additional credit to *reduce* a service member's tax liability. In contrast, a positive change to income tax or payroll tax means that there is an increase in a service member's tax liability. As described in Chapter Three, contributing to an FSA can have a positive or negative impact on a service member's tax liability depending on his/her family's individual cir-

TABLE 4.1
2020 Basic Pay Rate Assumptions

Paygrade	Most-Common Years of Service	Basic Pay Rate
E1	< 2	$20,797.00
E2	< 2	$23,310.00
E3	< 2	$24,512.40
E4	3	$30,085.20
E5	4	$34,696.80
E6	8	$43,837.20
E7	18	$58,708.80
E8	18	$64,735.20
O1	< 2	$39,445.20
O2	3	$59,612.40
O3	4	$70,167.60
O4	10	$89,524.80
O5	18	$111,330.00
O6	24	$137,613.60

SOURCE: Authors' calculations using Active Duty Master File for September 2021 provided by the Defense Manpower Data Center (DMDC) and the 2020 basic pay table from DoD (2020).

cumstances. When the total tax change is positive—that is, the service member is estimated to experience a tax increase from contributing to an FSA—we highlight the estimate in red font.

The second viewpoint focuses on estimating the change in tax liability because of contributions to an FSA by family earnings. Considering the results from the standpoint of earnings has two advantages. It allows us to recognize that income for a member might vary from others in the same paygrade because of the receipt of special and incentive pay, such as an aviator bonus (up $35,000 per year for some specialties) or sea pay. Second, it allows us to recognize that the working spouses of married service members might earn substantially more than the assumed $25,000 in the paygrade analysis, affecting the extent of the tax change due to contributing to an FSA. For this analysis, we varied earnings by $1,000 bins from $20,000 to $200,000 for unmarried service members and married service members with a nonworking spouse. For married service members with a working spouse, we varied earnings by $1,000 from $30,000 to $200,000. We then plot the change in total taxes from contributing to an FSA by earnings to understand whether certain income groups would be adversely or positively affected by participating in an FSA and determine the range of the total tax change.

The results are presented as a series of figures that show how the tax change from contributing to an FSA varies with earnings. These figures plot the results for members with zero, one, or two dependents. We also conducted the simulations for members with three dependents, and those results are similar with those for members with two dependents.

Estimated Tax Benefit from DCFSA

The population who would potentially benefit from a DCFSA are service members with children under age 13.[4] Using 2020 DMDC Active Duty Pay File and Defense Enrollment Eligibility Reporting System Point-in-Time Extract (DEERS PITE) data, we find that 35 percent of enlisted service members and 43 percent of officers have children under age 13 (see Table 4.2). These percentages represent upper bounds for the population of service members who could benefit from a DCFSA because the spouse must earn at least the amount contributed to the DCFSA for married service members. Fifty-four percent and 65 percent of enlisted service members and officers with children under age 13, respectively, are married. Among unmarried service members with children under age 13, it is most common to have one child for both enlisted service members and officers. For married service members with children under age 13, it is most common to have one child among enlisted members and two children among officers.

As we discussed in Chapter Three, among this population, only members with expenses for dependent care that is provided by installation-based facilities or by off-base community

TABLE 4.2

Percentage of Service Members by Marital Status and Number of Children Under Age 13

	Enlisted	Officer
Unmarried, one child	9%	7%
Unmarried, two children	5%	5%
Unmarried, three-plus children	2%	2%
Married, one child	8%	10%
Married, two children	7%	11%
Married, three-plus children	4%	7%
Any children	35%	43%

SOURCE: Authors' calculations using 2020 Active Duty Pay File and DEERS PITE data provided by the DMDC.

[4] Our analysis focuses on service members with children under age 13. However, service members with other dependents (e.g., disabled adult dependents) may also be eligible for a DCFSA. We were unable to find data on the number of service members with dependents age 13 and older who would qualify for a DCFSA.

care where the member did not receive fee assistance would benefit from a DCFSA. Although Table 4.2 provides information on the percentage of members with different numbers of children under age 13 by marital status, we lack the data to further decompose the figures in Table 4.2 by whether they received off-base care without the subsidy and, therefore, would benefit from a DCFSA.[5]

Tax Benefit from Contributing to a DCFSA by Paygrade

To understand which service members might benefit from participating in a DCFSA, we first examine the distribution of service members with children across enlisted and officer paygrades. Tables 4.3 and 4.4 show that the bulk of service members with children are in certain paygrades. Among enlisted service members with children under age 13, 87 percent of enlisted members are in E4 to E7 paygrades (Table 4.3). Among officers with children under age 13, 86 percent are in the O3 to O5 paygrades (Table 4.4). As a result, our analysis of the benefit of DCFSAs by paygrade will be limited to the E4–E7 and O3–O5 paygrades. As noted earlier, we are unable to further decompose the tabulations in Tables 4.3 and 4.4 by whether these members received installation-based care or off-base care without a subsidy because only those in these groups would likely benefit from a DCFSA.

TABLE 4.3

Distribution of Active-Duty Enlisted Service Members with Children Under Age 13 Across Paygrades in 2020

Paygrade	With Children Under 6	With Children Ages 6–12	With Any Children Under 13
E1	0%	0%	0%
E2	1%	0%	1%
E3	5%	2%	4%
E4	18%	6%	14%
E5	29%	18%	24%
E6	29%	36%	30%
E7	14%	28%	19%
E8	14%	28%	19%
E9	1%	2%	1%

SOURCE: Authors' calculations using 2020 Active Duty Pay File and DEERS PITE data provided by the DMDC.

[5] DoD conducts an annual Status of Forces Survey of Active Members and a survey of military spouses that asks respondents about whether they receive off-base or on-base child care. However, the surveys do not ask respondents who receive off-base care about whether they are receiving fee assistance.

TABLE 4.4
Distribution of Active-Duty Officers with Children Under Age 13 Across Paygrades in 2020

Paygrade	With Children Under Age 6	With Children Ages 6–12	With Any Children Under Age 13
O1	4%	2%	3%
O2	7%	4%	6%
O3	37%	21%	30%
O4	36%	35%	34%
O5	15%	30%	22%
O6	2%	8%	6%
O7	0%	0%	0%
O8	0%	0%	0%
O9	0%	0%	0%
O10	0%	0%	0%

SOURCE: Authors' calculations using 2020 Active Duty Pay File and DEERS PITE data provided by the DMDC.

We first provide several specific examples of how contributing $5,000 to a DCFSA affects total taxes to illustrate how FSA contributions affect different components of total tax. By *total tax* we mean the change in income taxes due to the FSA contribution *plus* the change in payroll tax because the tax base upon which the member's Social Security and Medicare tax rates are applied is lower when the member contributes to an FSA. The examples also illustrate how contributing to a DCFSA has the potential to increase and decrease total tax depending on a service member's specific circumstances. Chapter Two reviewed the different tax component, and we broadly discussed the interactions between FSA contributions and benefits from these other components in Chapter Three. Here, we provide some specific examples, given our methodology for estimating tax change.

Table 4.5 reports the changes in tax before credits, CDCTC, Child Tax Credit, EITC, total income tax, payroll tax, and total tax (i.e., the sum of total income tax and payroll tax) for an unmarried E4 with one child, a married E6 with two children, and a married O4 with two children. As described earlier, we assume service members in each paygrade earn the basic pay rate for the most-common observed years of service for that paygrade based on 2021 data. And we assume military spouses earn $25,000. The table reports tax changes in each tax component. As noted earlier in this chapter, depending on the tax component, a positive change represents a tax increase or tax decrease. Positive changes in tax before credits, total income tax, payroll taxes, and total taxes are tax increases, whereas positive changes in CDCTC, Child Tax Credit, and EITC are tax decreases. In other words, increases in tax credits *reduce* tax liability.

TABLE 4.5
Changes in Taxes from Contributing $5,000 to DCFSA by Tax Component, 2020

Tax Component	Unmarried E4 with One Child	Married E6 with Two Children	Married O5 with Two Children
Income Tax			
Tax before credits	–$500.00	–$600.00	–$1,100.00
Child Tax Credit	+$280.20	$0.00	$0.00
CDCTC	–$823.72	–$1,000.00	–$1,000.00
EITC	+$799.00	$0.00	$0.00
Total income tax	–$755.48	+$400.00	–$100.00
Payroll tax	–$382.50	–$382.50	–$382.50
Total tax	–$1,137.98	+$17.50	–$482.50

For the unmarried E4 with one child, we estimate that tax before credits decreases by $500 when the member contributes $5,000 to a DCFSA. The magnitude of the decrease in tax before credits depends on the service member's tax bracket. The Child Tax Credit and EITC both increase by $280.20 and $799, respectively, because the reduction in taxable earnings causes the service member to be eligible for larger refundable credits. Because contributions to the DCFSA reduce eligible expenses for the CDCTC dollar for dollar, the CDCTC will decrease. For the unmarried E4 with one child, we estimate that the CDCTC decreases by $823.72. In sum, the total change in income tax is –$755.48 (= $823.72 – $500.00 – $280.20 – $799.00). As mentioned, payroll taxes for the member also decrease because the FSA contribution reduces the amount of earnings subject to Social Security and Medicare taxes. Adding in the payroll tax change of –$382.50 yields a total tax change of –$1,137.98 for the E4, or a tax savings in this case.

For a married E6 with two children, we estimate that a $5,000 contribution to a DCFSA would reduce tax before credits by $600 and reduce the CDCTC by $1,000. The Child Tax Credit and EITC are both unaffected by the DCFSA contribution. The DCFSA contribution does not affect the amount of Child Tax Credit that can be claimed because the married E6's family earnings are well beyond the phase-in range of the Child Tax Credit. The EITC is unaffected because the service member's family earnings are well beyond the EITC eligibility range and the family is unable to claim the EITC with or without the DCFSA contribution. The total change in income tax is +$400. The total tax change after accounting for the payroll tax reduction is +$17.50 (in red font), demonstrating that certain service members could experience a tax increase from contributing to a DCFSA, albeit small in this case.

For a married O5 with two children, we estimate that a $5,000 contribution to a DCFSA would reduce tax before credits by $1,100 and reduce the CDCTC by $1,000. Similar with the case of a married E6 with two children, the Child Tax Credit and EITC are unaffected by the DCFSA contribution for the same reasons described for the E6. However, unlike the married E6, the married O5 is estimated to experience a tax savings from contributing to a DCFSA because the reduction in tax before credits is larger than the reduction in CDCTC with a total change in income tax of –$100. Including the payroll tax savings increases the total savings to $482.50.

These examples demonstrate that not all service members would necessarily benefit from contributing to a DCFSA and some may actually experience a small cost, even if the care they receive is installation-based or off-base and unsubsidized. However, members who would experience a tax increase may choose not to participate in the DCFSA, which would prevent them from experiencing this cost. We investigate this further by presenting the estimated change in total taxes from contributing $5,000 to a DCFSA by marital status, paygrade, and number of children in Tables 4.6 and 4.7 with Table 4.6 showing the results for *married* service members and Table 4.7 showing the results for *unmarried* members. We find that service members with one child are estimated to experience a total tax decrease from contributing $5,000 to a DCFSA across all paygrades investigated and regardless of marital status. For

TABLE 4.6

Change in Taxes from Contributing $5,000 to DCFSA, Unmarried Service Members, by Paygrade and Number of Children, 2020

Paygrade	One Child	Two Children	Three Children
E4	–$1,137.98	–$1,435.50	–$685.50
E5	–$965.89	–$1,435.50	–$1,315.02
E6	–$831.81	–$1,006.43	–$1,095.04
E7	–$382.5	+$17.50	+$17.50
03	–$382.5	+$17.50	+$17.50
04	–$382.5	–$482.50	–$482.50
05	–$382.5	–$582.50	–$582.50

TABLE 4.7

Change in Taxes from Contributing $5,000 to DCFSA, Married Service Members, by Paygrade and Number of Children, 2020

Paygrade	One Child	Two Children	Three Children
E4	–$382.50	–$655.88	–$1,035.50
E5	–$382.50	+$17.50	–$434.69
E6	–$382.50	+$17.50	+$17.50
E7	–$382.50	+$17.50	+$17.50
03	–$382.50	+$17.50	+$17.50
04	–$882.50	–$482.50	–$482.50
05	–$882.50	–$482.50	–$482.50

service members with one child who are either unmarried in the grades of E7 or O3 to O5 or who are married (any grade), the change in their income tax from contributing $5,000 to a DCFSA is zero. But the total tax change from contributing $5,000 to an FSA is –$382.50 for these individuals because the payroll tax reduction is $382.50. For unmarried service members with one child in the E4 to E6 paygrades, we estimate a reduction in taxes from contributing $5,000 to an FSA ranging from $831.81 to $1,137.98.

For unmarried enlisted service members with two or three children, we estimate a tax savings from contributing to a DCFSA ranging from $1,006.43 to $1,435.50 for those in the E4 to E6 paygrades. However, service members with two or three children in certain paygrades are predicted to experience a small tax increase of $17.50—specifically, unmarried E7s and O3s, married E6s, E7s, and O3s, and married E5s with two children. These members experience a $400 increase in income tax liability that is partially offset by the $382.50 reduction in payroll taxes as shown for married E6s with two children in Table 4.5. Service members with two or three children in the O4 and O5 paygrades and married E5s with three children experience a tax savings from contributing $5,000 to a DCFSA ranging from $434.69 to $582.50.

Benefit from Contributing to a DCFSA by Earnings

The estimates for the total tax change from contributing $5,000 to a DCFSA by paygrade suggest that there are certain income ranges in which a DCFSA would not be advantageous to the service member. To investigate this further, we estimate the change in taxes from contributing $5,000 to a DCFSA at different earnings to understand how the tax change from contributing to a DCFSA varies. Figures 4.1 and 4.2 present the results for unmarried and married service members, respectively. The x-axis in these figures present gross family earnings, meaning these are earnings without the $5,000 reduction from the DCFSA contribution. For married service members, we assume that the military spouse earns at least $5,000, so the couple qualifies for the maximum DCFSA contribution. We used NBER TAXSIM to estimate the change in federal income tax from contributing $5,000 to a DCFSA and added in the estimated payroll tax reduction to calculate the total tax change.

Figures 4.1 and 4.2 show who gains and who loses from contributing $5,000 to a DCFSA by gross earnings. Those who benefit experience a negative change in taxes, and those who lose experience a positive change in taxes. The figures show the complex, nonlinear nature of the U.S. tax code with the tax change rising and falling at different earnings levels as certain credits phase in and out and as tax brackets change. In general, for lower levels of earnings, the tax benefit from contributing to a DCFSA is driven by increases in the amount of EITC that can be claimed, and to a smaller extent, increases in the amount of the refundable Child Tax Credit (with the exception of the lowest earnings groups as noted here). As the EITC phases out at higher levels of earnings, the tax benefit from contributing $5,000 to a DCFSA decreases. Once the EITC phases out (i.e., earnings exceed the means tested threshold to claim the EITC), the tax benefit from contributing to a DCFSA plateaus until earnings reach a level where the service member moves up to the next tax bracket and the reduction in

FIGURE 4.1

Change in Taxes by Gross Earnings from Contributing $5,000 to DCFSA, Unmarried Service Members, 2020

NOTE: Gross earnings are equal to family earnings without the FSA contribution deducted.

FIGURE 4.2

Change in Taxes by Gross Earnings from Contributing $5,000 to DCFSA, Married Service Members, 2020

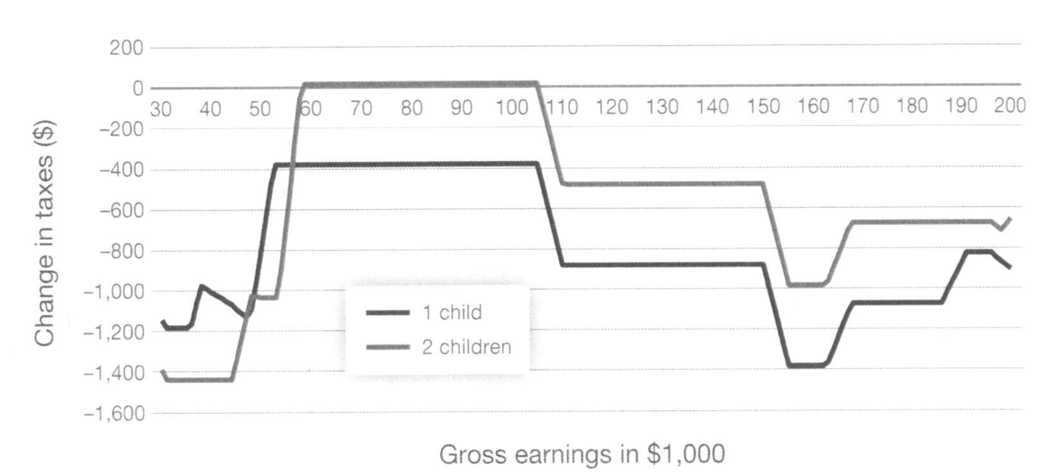

NOTE: Gross earnings are equal to family earnings without the FSA contribution deducted.

income tax (before credits) increases. Additional nonlinearities in the U.S. tax code cause the tax benefit from contributing to a DCFSA to increase and decrease at higher earnings levels, as shown in Figures 4.1 and 4.2.

At most earnings levels, service members would experience a tax benefit (i.e., a negative change in total taxes), but there are certain earnings ranges where service members are estimated to experience a tax increase. We note that members who would experience a tax increase could choose not to participate in the DCFSA, causing their tax change to be zero. Unmarried service members with one child are estimated to experience a tax benefit (i.e., a negative change in total taxes) at all earnings levels we considered with the total tax reduction ranging from –$382.50 to –$1,482.50. For unmarried service members with two children, we find a tax benefit (meaning a reduction in total taxes) at most earnings levels except for those with earnings between $20,000 and $22,000 and between $53,000 and $72,000. For unmarried service members with two children and earnings between $20,000 and $22,000, the total tax change is positive (there is no tax benefit) from contributing $5,000 to a DCFSA because the FSA contribution reduces the amount of the refundable Child Tax Credit that can be claimed and this reduction in credit is not offset by the estimated increase in EITC or estimated reductions in tax before credits and payroll taxes. For this group, the tax increase is at most $226. We note that there are few service members with earnings in this range; based on 2020 DMDC data, we estimate that there are 536 active-duty service members with earnings between $20,000 and $22,000. Unmarried service members between $53,000 and $72,000 experience a small total tax increase of $17.50 from contributing $5,000 to a DCFSA. This tax increase is due to the loss in CDCTC, which is not offset by a reduction in tax before credits and payroll taxes. We estimate that 14,304 service members have earnings between $53,000 and $72,000. Once earnings exceed $72,000, the additional reduction in tax before credits offsets the loss in CDCTC and the total tax change turns negative (meaning the tax benefit is positive).

Like unmarried service members with one child, married service members with one child are estimated to experience a tax benefit from contributing $5,000 to a DCFSA at all earnings levels, assuming they have eligible child care expenses. For these members, the tax decrease ranges from $382.50 to $1,382.50. Married service members with two children and earnings between $59,000 and $105,000 are estimated to experience a small tax increase of $17.50 from contributing $5,000 to a DCFSA. Similar with their unmarried counterparts, the tax increase is due to the loss in CDCTC, which exceeds the reduction in tax before credits and payroll taxes. For all other earnings levels we considered, married service members with two children are estimated to have a tax benefit from contributing to a DCFSA, with the tax reduction ranging from $52 to $1,435.50.

Benefit from Health Care Flexible Spending Accounts

We estimate the total tax change from contributing $500 and $2,850 to an HCFSA to investigate the potential benefit to service members from participating in an HCFSA. We present results first by paygrade and second by earnings. As discussed in Chapter Three and in the introduction to this chapter, active members enrolled in TRICARE Prime have virtually zero out-of-pocket health care expenses that would be eligible for an HCFSA, while family members enrolled in TRICARE Prime have average expenses of about $100 (although they may have other HCFSA eligible expenses). Average expenses for a military family enrolled in TRICARE Select are $500. Thus, the estimates we show in this subsection are more relevant to the 17 percent of family members covered by TRICARE Select who either experience the average amount of out-of-pocket expense of $500 or who contribute the maximum to the HCFSA to cover predictable and larger expenses.

Benefit from Contributing to an HCFSA by Paygrade

We calculated the total change in taxes from contributing to an HCFSA by marital status, paygrade, number of children, and additionally by spouse work status for married service members. Before discussing the results by paygrade, we first present in Table 4.8 several examples showing the total tax change from contributing $500 to an HCFSA among service members in the E5 paygrade to demonstrate how contributing to an HCFSA can affect the different tax components. Similar to the DCFSA examples in Table 4.5, a positive change in Table 4.8 represents a tax increase or decrease depending on the tax component. Positive changes in tax credits represent reductions in tax liability while positive changes in tax before credits, total income tax, payroll taxes, and total taxes represent increases in tax liability.

TABLE 4.8

Changes in Tax Components from Contributing $500 to HCFSA Among Service Members in the E5 Paygrade, 2020

Tax Component (1)	Unmarried E5 with No Children (2)	Married E5 with Nonworking Spouse and Three Children (3)	Married E5 with Working Spouse and Two Children (4)
Income Tax			
Tax before credits	-$60	-$50	-$60
Child Tax Credit	$0	-$50	$0
CDCTC	$0	$0	$0
EITC	$0	+$105.30	$0
Total	-$60	-$105.30	-$60
Payroll tax	-$38.25	-$38.25	-$38.25
Total tax	-$98.25	-$143.55	-$98.25

NOTES: The numbers in columns (2) and (4) can be added to obtain ***Total tax***, with the signs on the individual numbers preserved. The numbers in column (3) cannot be added because reductions in taxes enter with a minus sign (a benefit) but ***reductions*** in tax credits also enter with a minus sign (a negative benefit). Therefore, they are not additive to total net benefits with preserved signs. In the case of columns (2) and (4), the effect on tax credits is zero, so all of the nonzero numbers are additive with signs preserved.

For unmarried E5s with no children, we estimate that income tax before credits would decrease by $60 and payroll taxes would decrease by $38.25, yielding a total tax decrease of $98.25. For a married E5 with a nonworking spouse and three children, contributing $500 to an HCFSA reduces income tax before credits and the CTC by $50. The CTC goes down with the HCFSA contribution because he or she has earnings that put the couple in the phase-in range of the CTC. Therefore, reducing earnings by the HCFSA contribution amount *decreases* the amount of CTC that can be claimed. In contrast, reducing earnings by the HCFSA contribution *increases* the amount of EITC that can be claimed by $105.30 because the service member moves left in the phase-out range of the EITC similar to the example shown in Figure 3.7. The overall income tax decrease is $105.30. Adding in the $38.25 reduction in payroll taxes results in a final total tax *decrease of* $143.55. For a married E5 with a working spouse who makes $25,000 and has two children, we estimate that income taxes before credits would decrease by $60, causing a reduction in income tax of $60. After accounting for the payroll tax reduction, the total tax savings from contributing $500 to an HCFSA is estimated to be $98.25 for this service member, same as the savings estimated for an unmarried E5 with no children.

The examples in Table 4.8 illustrate how the benefit from contributing to an HCFSA depends on the number of children, marital status, and work status of the spouse. It also illustrates how the decision to contribute to an HCFSA is complicated by how FSA contributions interact with other components of the U.S. tax code in a way that is not necessarily favorable to the taxpayer. To further investigate whether certain groups might experience a tax increase from contributing to an HCFSA, we present estimates of the total tax change by paygrade from contributing to an HCFSA by marital status, spouse work status, and number of children. Tables 4.9 through 4.11 contain the tax change estimates when we assume $500 is contributed to an HCFSA and Tables 4.12 through 4.14 contain the tax change estimates when we assume $2,850 is contributed to an HCFSA.[6]

We estimate that, in most cases, there would be a tax savings from contributing to an HCFSA for unmarried service members and married service members with a working spouse across the paygrades and for those with and without children. That is, in virtually all cases, the change in total tax is negative. The tax savings from contributing to an HCFSA is estimated to be greater when we assume a larger HCFSA contribution. For unmarried service members, we estimate tax reduction ranging from $68.55 to $208.25 from contributing $500 to an HCFSA and from $99.48 to $1,187.03 from contributing $2,850 to an HCFSA. For married service members with a working spouse, the tax reduction from contributing $500 and $2,850 to an HCFSA varies from $98.25 to $203.55 and from $527.97 to $1,160.24, respectively.

Married service members with a nonworking spouse with two children in the E1 paygrade and with three children in the E1, E2, and E3 paygrades are estimated to experience a tax

[6] For completeness, Tables 4.9 through 4.14 show estimated tax change for infrequent cases, such as for unmarried E1s with children. DoD Instruction 1304.26, the instruction that describes DoD enlistment policy, disallows unmarried individuals from enlisting if they have dependents under the age 18, without a waiver.

TABLE 4.9

Change in Taxes by Paygrade from Contributing $500 to HCFSA, Unmarried Service Members, by Paygrade and Number of Children, 2020

Paygrade	No Children	One Child	Two Children	Three Children
E1	–$88.25	–$118.15	–$68.55	–$68.55
E2	–$98.25	–$118.15	–$143.55	–$68.55
E3	–$98.25	–$118.15	–$143.55	–$68.55
E4	–$98.25	–$168.15	–$143.55	–$68.55
E5	–$98.25	–$178.15	–$203.55	–$143.55
E6	–$98.25	–$98.25	–$203.55	–$203.55
E7	–$148.25	–$98.25	–$98.25	–$98.25
O1	–$98.25	–$178.15	–$203.55	–$203.55
O2	–$148.25	–$98.25	–$98.25	–$98.25
O3	–$148.25	–$98.25	–$98.25	–$98.25
O4	–$173.25	–$148.25	–$148.25	–$148.25
O5	–$158.25	–$158.25	–$158.25	–$158.25
O6	–$158.25	–$183.25	–$208.25	–$208.25

increase from contributing to an HCFSA. These increases in total tax are driven by reductions in the CTC. For these service members, their earnings excluding the HCFSA contribution place them in the phase-in range of the CTC and further reductions in earnings from the HCFSA reduce the amount of CTC that can be claimed. The estimated **tax increase** for this group is $36.75 under the $500 contribution assumption and $209.48 under the $2,850 assumption. Although the simulations are meant to show the potential tax consequences of contributing to an HCFSA for all paygrades and family statuses, in practice, few would experience a tax increase. In the DMDC data for 2020, about 1,500 service members are married E1s with two children or married E1s, E2s, and E3s with three children, and likely a subset of this group would also have a nonworking spouse. E1s who are married with two children represent only 0.7 percent of E1s and E1s, E2s, and E3s who are married with three children represent 0.2 percent 0.3 percent, and 0.5 percent of their respective paygrade.

Benefit from Contributing to an HCFSA by Earnings

Similar to the results by paygrade, we estimate that service members across most earnings levels would experience a tax benefit from contributing to an HCFSA as shown in Figures 4.3 through 4.8. If we assume that service members contribute $500 to an HCFSA, then the tax benefit is small with a maximum tax savings of $208.25. The maximum tax benefit is larger,

TABLE 4.10

Change in Taxes by Paygrade from Contributing $500 to HCFSA, Married Service Members, Nonworking Spouse, by Paygrade and Number of Children, 2020

Paygrade	No Children	One Child	Two Children	Three Children
E1	–$76.50	–$38.25	$36.75	$36.75
E2	–$38.25	–$38.25	–$38.25	$36.75
E3	–$38.25	–$38.25	–$38.25	$36.75
E4	–$88.25	–$118.15	–$143.55	–$68.55
E5	–$88.25	–$168.15	–$143.55	–$143.55
E6	–$88.25	–$168.15	–$193.55	–$193.55
E7	–$98.25	–$98.25	–$98.25	–$98.25
O1	–$88.25	–$168.15	–$193.55	–$143.55
O2	–$98.25	–$98.25	–$98.25	–$98.25
O3	–$98.25	–$98.25	–$98.25	–$98.25
O4	–$98.25	–$98.25	–$98.25	–$98.25
O5	–$148.25	–$148.25	–$148.25	–$148.25
O6	–$148.25	–$148.25	–$148.25	–$148.25

TABLE 4.11

Change in Taxes by Paygrade from Contributing $500 to HCFSA, Married Service Members, Working Spouse, by Paygrade and Number of Children, 2020

Paygrade	No Children	One Child	Two Children	Three Children
E1	–$98.25	–$178.15	–$203.55	–$203.55
E2	–$98.25	–$98.25	–$203.55	–$203.55
E3	–$98.25	–$98.25	–$203.55	–$203.55
E4	–$98.25	–$98.25	–$98.25	–$203.55
E5	–$98.25	–$98.25	–$98.25	–$98.25
E6	–$98.25	–$98.25	–$98.25	–$98.25
E7	–$98.25	–$98.25	–$98.25	–$98.25
O1	–$98.25	–$98.25	–$98.25	–$98.25
O2	–$98.25	–$98.25	–$98.25	–$98.25
O3	–$98.25	–$98.25	–$98.25	–$98.25
O4	–$148.25	–$148.25	–$148.25	–$148.25

Table 4.11—Continued

Paygrade	No Children	One Child	Two Children	Three Children
O5	–$148.25	–$148.25	–$148.25	–$148.25
O6	–$198.25	–$198.25	–$198.25	–$198.25

TABLE 4.12

Change in Taxes by Paygrade from Contributing $2,850 to HCFSA, Unmarried Service Members, by Paygrade and Number of Children, 2020

Paygrade	No Children	One Child	Two Children	Three Children
E1	–$503.03	–$452.46	–$99.48	–$99.48
E2	–$523.73	–$673.46	–$712.24	–$390.74
E3	–$547.78	–$673.45	–$818.24	–$390.74
E4	–$560.03	–$958.46	–$818.24	–$390.74
E5	–$560.03	–$997.40	–$1,142.17	–$818.24
E6	–$560.03	–$683.21	–$1,160.24	–$1,160.24
E7	–$845.03	–$560.03	–$560.03	–$560.03
O1	–$560.03	–$1,015.46	–$1,160.23	–$1,160.24
O2	–$845.03	–$560.03	–$560.03	–$560.03
O3	–$845.03	–$560.03	–$560.03	–$560.03
O4	–$1,003.79	–$845.03	–$845.03	–$845.03
O5	–$902.03	–$902.03	–$902.03	–$902.03
O6	–$902.03	–$1,131.33	–$1,187.03	–$1,187.03

TABLE 4.13

Change in Taxes by Paygrade from Contributing $2,850 to HCFSA, Married Service Members, Nonworking Spouse, by Paygrade and Number of Children, 2020

Paygrade	No Children	One Child	Two Children	Three Children
E1	–$436.05	–$218.03	$209.48	$209.48
E2	–$313.86	–$218.03	–$112.03	$209.48
E3	–$221.88	–$218.03	–$218.03	$209.48
E4	–$503.03	–$673.45	–$818.24	–$390.74
E5	–$503.03	–$958.46	–$818.24	–$818.24
E6	–$503.03	–$958.46	–$1,103.24	–$921.95

Table 4.13—Continued

Paygrade	No Children	One Child	Two Children	Three Children
E7	–$560.03	–$560.03	–$560.03	–$767.49
O1	–$503.03	–$958.46	–$1,082.76	–$818.24
O2	–$560.03	–$560.03	–$560.03	–$577.19
O3	–$560.03	–$560.03	–$560.03	–$560.02
O4	–$560.03	–$560.03	–$560.03	–$560.03
O5	–$845.03	–$845.03	–$845.03	–$845.03
O6	–$845.03	–$845.03	–$845.03	–$845.03

TABLE 4.14

Change in Taxes by Paygrade from Contributing $2,850 to HCFSA, Married Service Members, Working Spouse, by Paygrade and Number of Children, 2020

Paygrade	No Children	One Child	Two Children	Three Children
E1	–$527.97	–$983.39	–$1,128.18	–$1,128.18
E2	–$560.03	–$909.68	–$1,160.24	–$1,160.24
E3	–$560.03	–$717.54	–$1,160.24	–$1,160.24
E4	–$560.03	–$560.03	–$790.62	–$1,160.24
E5	–$560.03	–$560.03	–$560.03	–$560.03
E6	–$560.03	–$560.03	–$560.03	–$560.03
E7	–$560.03	–$560.03	–$560.03	–$560.03
O1	–$560.03	–$560.03	–$560.03	–$560.03
O2	–$560.03	–$560.03	–$560.03	–$560.03
O3	–$560.03	–$560.03	–$560.02	–$560.03
O4	–$845.03	–$845.03	–$845.03	–$845.03
O5	–$845.03	–$845.03	–$845.03	–$845.03
O6	–$1,130.03	–$1,130.03	–$1,130.03	–$1,130.03

FIGURE 4.3

Change in Taxes by Gross Earnings from Contributing $500 to HCFSA, Unmarried Service Members, 2020

NOTE: Gross earnings are equal to family earnings without the FSA contribution deducted.

FIGURE 4.4

Change in Taxes by Gross Earnings from Contributing $500 to HCFSA, Married Service Members, Working Spouse, 2020

NOTE: Gross earnings are equal to family earnings without the FSA contribution deducted. For married service members with a working spouse, we varied gross earnings from $30,000 to $200,000.

FIGURE 4.5

Change in Taxes by Gross Earnings from Contributing $500 to HCFSA, Married Service Members, Nonworking Spouse, 2020

NOTE: Gross earnings are equal to family earnings without the FSA contribution deducted.

FIGURE 4.6

Change in Taxes by Gross Earnings from Contributing $2,850 to HCFSA, Unmarried Service Members, 2020

NOTE: Gross earnings are equal to family earnings without the FSA contribution deducted.

FIGURE 4.7

Change in Taxes by Gross Earnings from Contributing $2,850 to HCFSA, Married Service Members, Working Spouse, 2020

NOTE: Gross earnings are equal to family earnings without the FSA contribution deducted. For married service members with a working spouse, we varied gross earnings from $30,000 to $200,000.

FIGURE 4.8

Change in Taxes by Gross Earnings from Contributing $2,850 to HCFSA, Married Service Members, Nonworking Spouse, 2020

NOTE: Gross earnings are equal to family earnings without the FSA contribution deducted.

at $1,187.03, when we assume an HCFSA contribution of $2,850. Like the estimates shown by paygrade, the figures show that certain service members with two children and very low earnings (below $23,000) would experience a tax increase from contributing to an HCFSA. The HCFSA contribution reduces earnings used to calculate the CTC, causing those with earnings in the phase-in range of the CTC schedule to experience a reduction in credit that can be claimed and increasing taxes owed. However, less than 1 percent of service members have earnings below $23,000 and at least two children, suggesting that few members would be adversely affected by participating in an HCFSA.

Benefit from Contributing to Both FSAs

In the previous two sections, we estimated the tax benefit from contributing to a DCFSA and HCFSA separately. In practice, service members with children under age 13 may choose to participate in both FSAs if they have eligible child care and eligible health care expenses. In this section, we estimate the change in taxes from contributing to both FSAs under two scenarios. The first scenario assumes that $500 is contributed to an HCFSA, the average amount of expenses for members covered by TRICARE Select, and the second scenario assumes that $2,850 is contributed to an HCFSA, the maximum contribution allowed. Under both cases, we assume that $5,000 is contributed to a DCFSA. We present results first by paygrade, focusing on the E4–E6 and O3–O5 paygrades because most service members with children under age 13 are in these paygrades. Next, we show the benefit from contributing to both FSAs graphically by earnings. Results are presented separately by marital status. Because both spouses must work for a married couple to qualify for a DCFSA, we assume that spouses of married service members are working and earn $25,000.

Benefit from Contributing to Both FSAs by Paygrade

Tables 4.15 to 4.18 show that contributing to both FSAs yields an estimated tax benefit for all of the paygrades we considered. Among unmarried service members, we find that the tax savings ranges widely from $80.75 to $1,579.05 under a $500 HCFSA contribution assumption and from $542.23 to $2,253.74 under a $2,850 HCFSA contribution assumption. The tax savings also vary across paygrades among married couples. Assuming a $500 HCFSA contribution, the tax savings for married couples is as low as $80.75 and as high as $1,239.05. With a $2,850 HCFSA contribution, the tax savings for married couples ranges from $542.53 to $2,195.74.

Benefit from Contributing to Both FSAs by Earnings

Like the results on the benefit from contributing to only an HCFSA, we find that the change in taxes from contributing to both FSAs varies with earnings. This can be seen in Figures 4.9 through 4.12, which show the change in taxes by gross earnings for different contribution

TABLE 4.15

Change in Taxes by Paygrade from Contributing $5,000 to DCFSA and $500 to HCFSA, Unmarried Members, by Paygrade and Number of Children, 2020

Paygrade	One Child	Two Children	Three Children
E4	–$1,299.65	–$1,579.05	–$754.05
E5	–$1,134.04	–$1,579.05	–$1,383.57
E6	–$1,009.96	–$1,237.48	–$1,531.90
E7	–$480.75	–$80.75	–$80.75
O3	–$480.75	–$80.75	–$80.75
O4	–$1,030.75	–$630.75	–$630.75
O5	–$1,140.75	–$740.75	–$740.75

TABLE 4.16

Change in Taxes by Paygrade from Contributing $5,000 to DCFSA and $500 to HCFSA, Married Service Members, by Paygrade and Number of Children, 2020

Paygrade	One Child	Two Children	Three Children
E4	–$480.75	–$869.43	–$1,239.05
E5	–$480.75	–$80.75	–$638.24
E6	–$480.75	–$80.75	–$80.75
E7	–$480.75	–$80.75	–$80.75
O3	–$480.75	–$80.75	–$80.75
O4	–$1,030.75	–$630.75	–$630.75
O5	–$1,030.75	–$630.75	–$630.75

TABLE 4.17

Change in Taxes by Paygrade from Contributing $5,000 to DCFSA and $2,850 to HCFSA, Unmarried Members, by Paygrade and Number of Children, 2020

Paygrade	One Child	Two Children	Three Children
E4	–$1,854.96	–$2,253.74	–$1,076.24
E5	–$1,924.35	–$2,253.74	–$1,705.76
E6	–$1,847.27	–$2,205.92	–$2,253.74
E7	–$942.53	–$542.53	–$562.57
O3	–$942.53	–$542.53	–$542.53
O4	–$1,727.53	–$1,327.53	–$1,327.53
O5	–$1,871.13	–$1,471.13	–$1,471.13

TABLE 4.18

Change in Taxes by Paygrade from Contributing $5,000 to DCFSA and $2,850 to HCFSA, Married Service Members, by Paygrade and Number of Children, 2020

Paygrade	One Child	Two Children	Three Children
E4	–$1,008.50	–$1,826.12	–$2,195.74
E5	–$942.53	–$854.92	–$1,594.93
E6	–$942.53	–$542.53	–$542.53
E7	–$942.53	–$542.53	–$542.53
O3	–$942.53	–$542.53	–$542.53
O4	–$1,727.53	–$1,327.53	–$1,327.53
O5	–$1,727.53	–$1,327.53	–$1,327.53

levels to both FSAs by marital status and number of children. In general, service members experience a tax benefit; the exception is members with very low earnings and two children, who represent a very small share of the overall service member population. The tax benefit for unmarried service members with one child varies between $480.75 and $1,690.75 when a $500 HCFSA contribution is assumed and between $707.59 and $2,669.53 when a $2,850 HCFSA contribution is assumed. For married service members with one child, contributing $500 to an HCFSA and $5,000 to a DCFSA yields a tax savings of between $480.75 to $1,580.75. When we assume a higher HCFSA contribution of $2,850, then the tax benefit for married service members with one child ranges between $942.53 and $2,512.53. For unmarried service members with two children and earnings of at least $22,000 and married couples with two children, the benefit from contributing $500 to an HCFSA and $5,000 to a DCFSA varies from $80.75 up to $1,579.05. Assuming a higher HCFSA contribution of $2,850 increases the range of the tax benefit from participating in both FSAs—$542.53 to $2,253.74 for married couples with two children and $470.93 to $2,269.53 for unmarried service members with children and earnings of at least $23,000.

Summary

In this chapter, we estimated the change in taxes from contributing to FSAs to understand the implications of offering an FSA option for active-duty service members and how those implications vary by family characteristics. In general, we find that most service members would experience a tax savings while certain groups would experience a tax increase if they chose to participate. These groups are generally small in size and/or the estimated tax increase itself is small. Furthermore, those who are estimated to experience a tax increase would choose not to participate in an FSA if they are aware that doing so would make them worse off.

FIGURE 4.9
Change in Taxes by Gross Earnings from Contributing $5,000 to DCFSA and $500 to HCFSA, Unmarried Service Members, 2020

NOTE: Gross earnings are equal to family earnings without the FSA contribution deducted.

FIGURE 4.10
Change in Taxes by Gross Earnings from Contributing $5,000 to DCFSA and $500 to HCFSA, Married Service Members, 2020

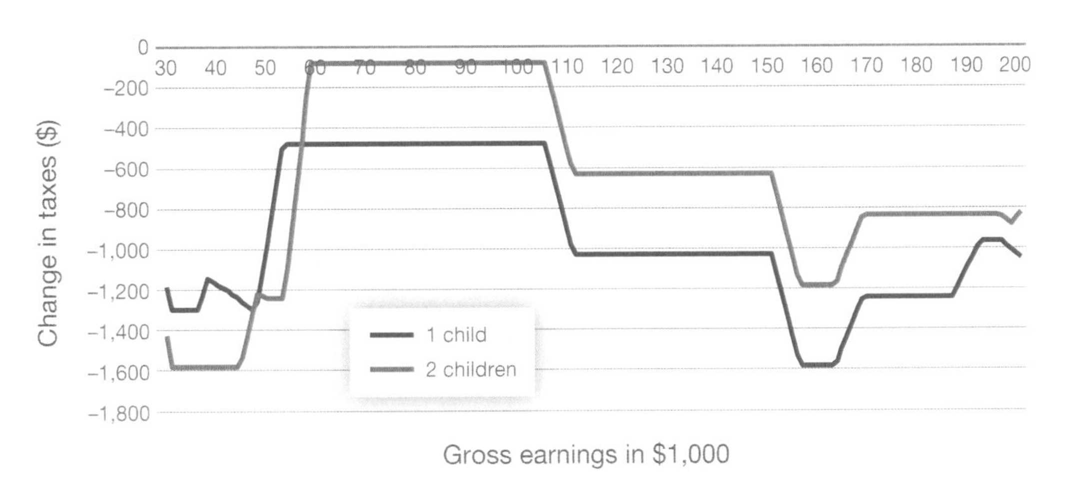

NOTE: Gross earnings are equal to family earnings without the FSA contribution deducted.

For the DCFSA, service members with one child, regardless of marital status, are estimated to experience a tax benefit from contributing to the FSA across earnings levels and paygrades investigated with the benefit estimated to be as high as $1,482.50. For service members with two or more children, most would experience a tax benefit from contributing $5,000 to a DCFSA with the benefit estimated to be as high as $1,435.50. There is a small

FIGURE 4.11
Change in Taxes by Gross Earnings from Contributing $5,000 to DCFSA and $2,850 to HCFSA, Unmarried Service Members, 2020

NOTE: Gross earnings are equal to family earnings without the FSA contribution deducted.

FIGURE 4.12
Change in Taxes from Contributing $5,000 to DCFSA and $2,850 to HCFSA, Married Service Members, 2020

NOTE: Gross earnings are equal to family earnings without the FSA contribution deducted.

group of unmarried service members with two children and very low earnings (between $20,000 and $22,000) who are estimated to experience a tax increase as high as $226 from contributing $5,000 to a DCFSA. In addition, for those with two children, unmarried service members with earnings between $53,000 and $72,000 and married service members with earnings between $59,000 and $105,000 are estimated to have a small increase ($17.50) in total taxes from contributing $5,000 to a DCFSA. Our analysis of the tax change by paygrade reveals the following groups would have earnings levels that would cause them to experience a small tax increase when they contribute $5,000 to a DCFSA: unmarried E7s and O3s with two or three children; married E5s to E7s and O3s with two children; and married E6s and E7s and O3s with three children.

When we examine the impact of contributing to the HCFSA and the impact of contributing to both FSAs, we find that the majority of service members would experience a tax savings from participating except for small groups of select married service members with very low earnings. For example, married service members with a nonworking spouse with two children in the E1 paygrade and with three children in the E1, E2, and E3 paygrades are estimated to experience a tax increase from contributing to an HCFSA. For this group, the tax increase ranges from $36.70 to $209.48 for a $500 and $2,850 HCFSA contribution, respectively. For those estimated to experience a tax benefit, the size of the benefit can be small or sizable, depending on the contribution amount assumed, whether a member participates in one or both FSAs, the demographics of the member, and family earnings. For example, service members who only participate in an HCFSA and contribute $500 to the account are estimated to experience a maximum tax savings of $208.25. In contrast, unmarried service members with one child who contribute $5,000 to a DCFSA and $2,850 to an HCFSA are estimated to experience a maximum tax savings of $2,669.53.

CHAPTER FIVE

FSA Costs and Savings to DoD and Total Savings to Members

This chapter considers the extent to which FSAs increase or decrease DoD personnel costs. On the one hand, FSAs would increase costs to DoD because of the recurring cost of administering an FSA program. On the other hand, FSAs would decrease DoD's payroll costs, meaning DoD's share of Social Security and Medicare taxes, just as it decreased member's payroll taxes.[1] The chapter also considers the costs to DoD of introducing and implementing an FSA program for service members, including setting up and adapting software and providing training and/or counseling to service members. Because such an FSA program has yet to be implemented for service members, we held discussions with federal government SMEs to gain insight into the general scale of these costs and what factors should be considered in estimating implementation costs. We also spoke with SMEs regarding the recurring cost of administering an FSA program.

In addition to the assessment of cost and savings to DoD, we provide estimates of the aggregate savings to service members across the active-duty force drawing on the estimates from Chapter Four under different assumptions about the take-up or participation rate of members. Together these calculations allow us to provide estimates of the potential overall cost and savings of introducing a DCFSA and HCFSA program for service members, accounting for the impact on both service members and DoD.

The chapter begins with a discussion of the specific sources of costs and savings to DoD and how we estimate them and also how we estimate aggregate savings, including the savings to service members. We then present estimates for DCFSAs and then for HCFSA. We conclude with brief summary of the chapter's findings.

[1] It is possible that the implementation of DCFSAs could change the composition of the force by increasing the number of service members with dependents. This type of composition change would increase personnel costs because certain benefits are more generous and costly for families (e.g., housing and subsistence allowances, health insurance).

Cost and Savings Categories

To understand the potential cost or savings to DoD for instituting FSAs for active-duty service members, we consider the four categories of costs and savings. The first category is the agency fee, the annual cost to administer the FSA program day to day. The fee is paid to the third-party vendor that would handle the FSA account (i.e., receive funds from service members and manage disbursements of funds for eligible expenses). To estimate this cost, we assume that were FSAs provided to active-duty service members, then OPM would be the administrator of the FSA program for members. OPM administers the FSA program for federal civilians (the Federal Flexible Spending Account Program [FSAFEDS]) and oversees the operations of the vendor (Health Equity).[2] Using a common administrator for federal employees and military members has a precedent; the Federal Retirement Thrift Investment Board[3] administers the Thrift Savings Plan (TSP) for both federal employees and military personnel and is able to recognize rules governing the TSP that are specific to each group.

We use published information on the fees charged to participate in FSAFEDS to estimate the annual FSA program costs to DoD. We assume that the FSA program would go into effect in 2025 and that the monthly fee charged to DoD for each FSA account would be $2.99 for HCFSAs and $2.99 for DCFSAs in 2025; in addition, OPM would charge a risk reserve fee of $0.25 per HCFSA per month that would go to OPM. The risk reserve fee also covers the cost to OPM of administering the FSA program (OPM, 2021).[4] Over 12 months, the total annual cost would be $38.88 and $35.88 per enrollee for an HCFSA and a DCFSA, respectively. According to SMEs, the agency fees charged for FSAFEDS are discounted after the plan initiation because enough funds in reserve accumulate to handle such situations as overpayments or large medical expenses occurring in the beginning of the calendar year. In theory, sufficient reserves would potentially be built up after the first year that FSAs were offered to active-duty service members. As a result, for each subsequent year after the first year that FSAs are implemented, we assume that the monthly cost per enrollee would decrease to $1.47 (= $1.22 + $0.25) for HCFSAs and $1.22 for DCFSAs, which are the average discounted fees for years 2017 to 2020 for each respective account (OPM, 2018; OPM, 2020).

The second category of costs are the additional ongoing overhead costs to administer the FSA program. These are the ongoing administrative costs that would be in addition to the agency fees and risk reserve fees that cover the contractor and OPM administrative costs discussed earlier; these can be thought of as the lower bound of the true total ongoing adminis-

---

[2] OPM (undated).

[3] Federal Register (undated).

[4] The 2024 fee per month per enrollee is predicted to be $2.90 based on Benefits Administration Letter Number 21-801 from OPM (2021). In addition, a $0.25 risk reserve fee is charged for each HCFSA to ensure enough funds are available to handle HCFSA reimbursements that are greater than existing contributions. The annual administration fee is predicted to grow by 3 percent annually between 2022 and 2024. We inflate the 2024 fee of $2.90 by 3 percent to estimate the 2025 fee of $2.99.

trative costs. The additional ongoing overhead costs include the costs from providing annual training about FSAs to active-duty service members, working groups to study the use and administration of FSAs, and any recurring payroll system costs associated with handling FSA deductions from active-duty service member paychecks. Because the FSA programs for military members are not currently offered, we could not obtain empirical estimates of ongoing overhead costs to DoD aside from the agency fees and risk reserve fees discussed earlier. Instead, we conducted a series of discussions with SMEs in DoD and OPM that touched on ongoing administrative costs and implementation challenges, sources of potential implementation costs, and administrative and/or legislative barriers to implementation.

The potentially most substantial ongoing overhead cost would be associated with providing annual financial literacy training about FSAs to active-duty service members, coordination among the services on the use and administration of FSAs, and any recurring payroll system costs associated with handling FSA deductions from active-duty service member paychecks. SMEs indicated that it seems likely comprehensive education would be required at initial entry training, along with condensed training at certain life milestones, such as marriage or the birth of a child. Training could also be required during the open enrollment season each year. One SME noted that a possible training strategy would be to require completion of a short course and an acknowledgement of the risks before individuals could enroll in an FSA program. According to one SME, this would be a similar strategy to what was used as part of the introduction of the Blended Retirement System.[5] However, based on our analysis in this chapter and our SME discussions, it seems likely that these ongoing costs would be less than the ongoing savings to DoD after the first year, and a fraction of the benefit to members in all years.

The third category of costs and savings is the savings to DoD from reduced payroll tax contributions.[6] As described in Chapter Four, for the majority of service members, DoD would save 7.65 percent of service member contributions to their FSAs. If a service member contributes the maximum of $5,000 to a DCFSA, then DoD would save $382.50. If a service member contributes the maximum of $2,850 to an HCFSA, then DoD would save $218.03. If a service member contributes $500 to an HCFSA, then DoD would save $38.25. For service members with earnings net of FSA contributions greater than the Social Security wage cap ($137,700 in 2020), DoD would save only the Medicare portion of payroll taxes or 1.45 percent of the FSA contribution because it would pay the maximum amount of Social Security taxes for that service member irrespective of whether he or she contributes to an FSA. We account for this in our estimates by assuming that the payroll tax savings for service members with earnings at or above $138,000 would equal 1.45 percent of the FSA contribution. We recognize that this is a rough approximation because an FSA contribution could reduce earnings

[5] The SMEs with whom we spoke were not aware of any detailed studies on the start-up costs associated with implementing FSAs.

[6] We note that the benefits to DoD of reduced payroll costs are an intergovernmental transfer and not real savings for the economy as a whole.

below the cap providing DoD with a 7.65 percent savings on a portion of the FSA contribution. Our tabulations using DMDC data reveal that 1.2 percent of service members had earnings of $138,000 or greater.

The fourth category of costs are the one-time implementation costs associated with the FSA program. Implementation costs include the costs associated with adapting payroll tax systems to manage pre-tax deductions in earnings for FSAs, developing the processes and contractual arrangements required to administer the plan, and creating training materials to provide to active-duty service members to educate them on how FSAs work. We consider these one-time implementation costs later in this chapter and revisit these costs in our discussion of an implementation plan in Chapter Six.

In our cost estimates, we are only able to quantify the agency fees and risk reserve fees charged to manage FSA accounts and the savings from reduced payroll taxes. We are not able to quantify the costs associated with implementation or other ongoing overhead costs. That said, we were able to gather some qualitative information on these costs from SMEs as we describe later in this chapter. Therefore, our estimates will underestimate the true cost of expanding FSAs to the active-duty population.

The magnitude of our cost estimates depends on the assumptions about the amount that service members would contribute to an FSA and the percentage of service members who would participate, i.e., the take-up rate. We present cost estimates using different assumptions about FSA contributions for HCFSAs and different assumptions about FSA participation for both types of FSAs to show the potential range of costs. For DCFSAs, we assume that participants would contribute the maximum of $5,000. The different assumed participation rates are applied uniformly, meaning that we assume the participation rate is the same across all earnings groups and across all demographic groups (i.e., by marital status and number of children). In practice, we might expect the participation rate to vary by these characteristics because the tax benefit from contributing to an FSA vary across these dimensions as we demonstrated in Chapter Four.

Along with the cost estimates, we calculate the total aggregate tax change to service members under the different contribution and participation rate assumptions. To determine the number of service members in each income bin, we supplemented the NBER TAXSIM analysis with tabulations of service members by the same $1,000 earnings bins using 2020 data from DEERS PITE and the DMDC Active Duty Pay File. We used service member federal taxable wages to create the service member counts by $1,000 earnings bins. For married couples with a working spouse, we assume that the spouse has $25,000 in earnings.[7] This means, for example, that a service member with $40,000 in federal taxable wages would be counted in the $65,000 income bin.

---

[7] We also estimated total tax savings assuming $50,000 in spouse earnings for married couples with a working spouse and the estimates vary by one percent or less to those presented in the main text, demonstrating that aggregate savings are not driven by our assumption that spouse earnings equal $25,000.

We make a few simplifying assumptions to calculate the aggregate tax change to use the total tax changes calculated by $1,000 earnings bins presented in Chapter Four. First, we assume that service members with more than three children would experience the same tax change as those with three children. Second, we assume that service members with at least $201,000 in earnings would experience the same tax change as those with $200,000 to $200,999 in earnings. Third, we exclude unmarried service members with gross earnings less than $20,000 and married service members with earnings less than $30,000 from the aggregate tax change calculation.

Recurring Savings Estimates for DCFSAs and HCFSAs

This section considers the recurring savings to DoD and to members in aggregate, first for DCFSA and then for HCFSA. Later in the chapter we consider implementation and ongoing overhead costs.

DCFSA Savings

To benefit from a DCFSA, a service member must have qualifying child care expenses for at least one child under age 13. Furthermore, for married service members, the military spouse must be working and earn at least the amount contributed to the DCFSA, or $5,000 given our assumed DCFSA contribution assumption. Using 2020 data from DMDC's Active Duty Pay File and DEERS PITE, we estimated that 436,544 active-duty service members had at least one child under age 13. Published tabulations from the 2019 Survey of Active Duty Spouses show that 64 percent of military spouses are in the labor force (DoD, 2020). If we assume that 64 percent of married active-duty service members have working spouse and assume these spouses earn at least $5,000, then the estimated number of service members eligible for a DCFSA falls to 348,900.[8] Although we predict that certain service members would experience a tax increase from contributing $5,000 to a DCFSA, we include them in our pool of service members eligible for a DCFSA because we do not observe child care arrangements or qualifying expenses of service members and instead vary the participation rate to show the full range of savings to DoD.[9]

---

[8] If the proportion of service members with children who are married changes, then the costs to DoD for providing a DCFSA to active-duty service members will also change. Fifty-six percent of service members with children in our extract are married. If instead we assume that 45 percent or 65 percent of service members with children are married, then the DoD cost estimates would decrease 4 percent and increase 5 percent, respectively.

[9] If we assume that service members who are estimated to have a tax decrease do not participate in a DCFSA and 100 percent of the remaining service members have $5,000 in qualifying expenses and choose to participate, then the total savings to DoD are estimated to be $91.9 million in the first year and $97.6 million in each subsequent year. In this scenario, the total annual tax savings to members who participate is estimated to be $206.3 million.

We estimate the savings to DoD from offering DCFSAs to active-duty service members as shown in Table 5.1. The savings in the first year of implementation range from $17,900,000 under a 15-percent participation rate up to $119,100,000 under a 100-percent participation rate. The savings are even larger in each subsequent year because the contractor administrative fee decreases after the first year. Specifically, the savings range from $19,000,000 with a 15-percent participation rate to $126,500,000 with a 100-percent participation rate.

In Table 5.1, we also include the total tax savings estimated for service members. The aggregate annual tax savings to service members from contributing $5,000 to a DCFSA ranges from $30.7 million under a 15-percent participation rate up to $204.9 million under a 100-percent participation rate. The average annual tax savings per service member from contributing $5,000 to a DCFSA is $587.19.

Although Table 5.1 shows a range of savings under different DCFSA participation rate assumptions, the upper range estimate assuming 100-percent participation is highly unlikely because only 64 percent of employed spouses with children 13 or younger routinely use child care according to the 2019 Survey of Active Duty Spouses. On the other hand, only members with children enrolled in installation-based programs or members using unsubsidized off-base child care would benefit from a DCFSA. In the 2019 Active Duty Spouse Survey, 37 percent of working spouses report using on-base care, implying a participation rate of 24 percent (37 percent × 64 percent of employed spouses using care). We have no data on the share of working spouses using child care that use an unsubsidized off-base provider. The spouse survey indicates that 73 percent of working spouses using child care use an off-base provider, but it is unclear what share of this 73 percent is unsubsidized and would therefore be eligible for a DCFSA. It is likely that some of 73 percent is unsubsidized, given that only 9 percent to 16 percent of children enrolled in military-supported child care were in off-base fee-assisted care, meaning that the DCFSA participation rate might likely be greater than 24 percent. Given that some share could benefit from a DCFSA because they receive unsubsidized care,

TABLE 5.1

DCFSA Savings to DoD and Total Tax Savings to Active Service Members, Assuming a $5,000 DCFSA Contribution for Each Participating Member

	FSA Participation Rate			
	15%	25%	50%	100%
DoD savings				
Total savings, year 1	$17,900,000	$29,800,000	$59,500,000	$119,100,000
Total savings, year 2+	$19,000,000	$31,600,000	$63,300,000	$126,500,000
Total annual tax savings to service members	$30,700,000	$51,200,000	$102,400,000	$204,900,000

NOTES: DoD estimates include annual administrative costs and payroll tax savings. DoD estimates do not include implementation costs or ongoing overhead costs.

we believe a participation rate above 25 percent—and perhaps as high as 50 percent—is the most reasonable estimate.[10]

HCFSA Savings

We assume that the population of active-duty service members are all eligible to participate in an HCFSA if they have eligible health care expenses. Based on 2020 data from DMDC Active Duty Pay File and DEERS PITE, there are a total of 1,189,753 active-duty service members. We calculate DoD costs under two different assumptions about the amount contributed to an HCFSA: a $500 contribution, which is the average out-of-pocket expenses for TRICARE Select members, and a $2,850 contribution, which the maximum allowed under an HCFSA. We also vary the participation from as low as 15 percent, which is approximately equal to the 17 percent of active-duty families covered by TRICARE Select, up to 100 percent to show the potential range of cost to DoD.

Except for the first year of implementation under a $500 assumed HCFSA contribution, we estimate a savings to DoD from implementing an HCFSA as shown in Table 5.2. Assuming a $500 average contribution among HCFSA participants, DoD incurs a cost of $0.63 per enrollee in the year of implementation and saves $20.61 per enrollee in each subsequent year.[11] Assuming that HCFSA participants contribute the maximum of $2,850, DoD saves $179.15 per enrollee in the year of implementation and saves $200.39 per enrollee in each subsequent year. When we assume a 15-percent participation rate, the total annual cost in the first year of implementation is $181,000 under the $500 contribution assumption. Assuming a $500 HCFSA contribution, the first year costs to DoD increase with the participation rate from $302,000 with a 25-percent participation rate up to $1,207,000 with a 100-percent participation rate. For each subsequent year, DoD would experience a savings from offering an HCFSA ranging from $3.6 million with a 15-percent participation rate up to $24.1 million with a 100-percent participation rate. Thus, the first-year DoD cost from offering an HCFSA when a $500 contribution is assumed is small relative to the savings estimated in the subsequent years, and DoD would experience a net savings two years after implementation. If instead participants contribute the maximum of $2,850, we estimate that DoD would experience a savings in the first year of implementation and each subsequent year. Specifically, the DoD savings would range from $31.6 million to $210.5 million in the first year and from $35.4 million to $235.8 million in each subsequent year.

---

[10] Future surveys of active-duty spouses should ask spouses about receipt of unsubsidized versus subsidized child care to assist in the development of better estimates of the share of spouses who could benefit from a DCFSA.

[11] The sum of the monthly agency fee and risk reserve fee in the first year is $3.24 per enrollee, which amounts to $38.88 per year per enrollee. The annual payroll tax savings for a $500 contribution or 7.65 percent of $500 or $38.25. The net change per enrollee is ($38.25 – $38.88), yielding a cost of $0.63. In the second year, the fees go down to $1.47 per enrollee or $17.64 per year per enrollee. The fees go down because a smaller fee suffices after reserves are built up in the FSA accounts. Net savings per enrollee are $20.61.

The aggregate tax savings to service members from contributing to HCFSAs are much larger in magnitude than the DoD savings estimated in Table 5.2. Although the payroll tax savings for DoD and service members are equivalent, the income tax savings to service members are much larger than the payroll tax savings, causing the aggregate tax change to service members to be orders of magnitude larger than the DoD savings. Under the $500 contribution assumption, the total tax savings to service members ranges from $19.8 million under a 15-percent participation rate to $132.3 million under a 100-percent participation rate. If we assume that participants contribute the maximum amount of $2,850, then service members' tax savings range from $112.7 million under the 15-percent participation rate up to $751.2 million under the full participation rate. The average benefit to service members is $111.19 and $631.38 when a $500 and $2,850 HCFSA contribution are assumed, respectively.

As mentioned, about 17 percent of military families participate in TRICARE Select, suggesting that the results for lowest assumed participation rate in the Table 5.2 is most relevant. On the other hand, TRICARE Prime members and their families might find the opportunity to cover over-the-counter medications with an FSA attractive. Furthermore, members and families might pursue medical and dental treatments that are currently not covered, such as orthodontia. Thus, the participation rate might be as high as 25 percent to 50 percent in Table 5.2.

TABLE 5.2
HCFSA Savings to DoD and Total Tax Savings to Active Service Members

	FSA Participation Rate			
	15%	25%	50%	100%
$500 contribution				
DoD savings				
Total savings year 1	–$181,000	–$302,000	–$604,000	–$1,207,000
Total savings year 2+	$3,600,000	$6,000,000	$12,000,000	$24,100,000
Total tax savings to service members	$19,800,000	$33,100,000	$66,100,000	$132,300,000
$2,850 contribution				
DoD savings				
Total savings year 1	$31,600,000	$52,600,000	$105,300,000	$210,500,000
Total savings year 2+	$35,400,000	$59,000,000	$117,900,000	$235,800,000
Total annual tax savings to service members	$112,700,000	$187,800,000	$375,600,000	$751,200,000

NOTES: DoD estimates include annual contractor administrative costs and payroll tax savings. DoD estimates do not include implementation costs or ongoing overhead costs. A negative DoD savings means that the DoD incurs a cost.

Implementation Costs

Because the FSA programs for military members are not currently offered, we could not estimate implementation costs empirically. Instead, as noted above, we conducted a series of discussions with SMEs in DoD and OPM regarding implementation challenges, sources of potential implementation costs, and administrative and/or legislative barriers to implementation. In this subsection, we summarize the findings from those discussions as they pertain to implementation costs.

The SMEs identified several sources of initial implementation costs. One would be to develop the processes and contractual arrangements required to administer the plan. Another would be the need to provide financial literacy training to members so they can best use the newly available benefits to meet their specific circumstances. Finally, the pay systems that would support the FSAs would need to be modified. Currently, three of the armed services are in the process of developing integrated pay and personnel systems (IPPSs) (i.e., the IPPS-Army [IPPS-A], the Air Force IPPS [AFIPPS], and the Navy's Navy Personnel and Pay System [NP2]) and are moving away from legacy systems (e.g., the Defense Finance and Accounting Service [DFAS]'s Defense Joint Military Pay System [DJMS]). Because each service is on its own timetable in the development of an IPPS, it is likely that both the legacy systems and the forthcoming systems (IPPS-A, AFIPPS, NP2) would require modification to support the FSA programs.[12]

With respect to the implementation costs related to plan administration, the SMEs noted that an FSA plan for service members could be administered by OPM as part of its Federal Flexible Benefits Plan (known as FedFlex) (OPM, 2021). However, doing so would require time to develop the processes and to plan how the FSA program for service members would be administered. One estimate would be about a year of planning that would be spent performing such tasks as modifying the OPM contract with its FSA vendor (Health Equity), allowing the vendor time to hire and train additional staff, and, depending on how many service members might enroll, allowing OPM to hire any needed additional staff. According to the SME discussions, administering a plan for service members would not be a fundamental departure from what currently occurs for the administration of the federal FSA program, unless there are changes in the legal parameters governing FSAs.

Another source of cost that the SMEs identified is the development of educational materials on FSAs for service members. As the discussion and analyses in the previous chapters make clear, the amount to elect for an HCFSA or a DCFSA is a complex decision, depending on a member's taxable income, unique family circumstances, and potentially uncertain future expenses. Despite the complexity of the decision facing members, SMEs stated that DoD could limit the cost of the development of the curriculum by taking advantage of many available off-the-shelf products.[13] The training would need to be adapted for each of the ser-

[12] See U.S. Army (undated); U.S. Air Force (undated); and U.S. Navy (undated).

[13] See, for example, Lorman Education Services (undated).

vices, but the core concepts would remain the same. If the FSA program for military personnel were to be rolled into FSAFEDS, then existing training materials for this federal program could be used, appropriately tailored to make it relevant to active-duty service members. SMEs noted that some care would need to be taken to avoid the appearance of advocating a particular course of action (i.e., marketing); DoD does not want to be criticized for encouraging a particular decision.

Several SMEs observed that adapting or creating the pay systems needed to support FSAs could prove to be a substantial challenge, both in terms of cost and time. Although the SMEs generally did not state specific costs or time scales, they noted that adapting the pay systems could involve months, half a year, or even substantially longer of government and contractor effort. Implementing FSAs on some legacy systems could prove to be particularly challenging, as active maintenance has slowed in anticipation of the transition to the IPPS systems. For the systems under development, full integration could result in slippage in their delivery schedule. The one cost estimate we heard was approximately $4 million dollars to modify a single service's forthcoming pay system to interact with an external module that implements an FSA program (similar with the way in which TSP for service members was implemented), and many times that to fully integrate FSA with that service's forthcoming pay system (that is, to implement the FSA program logic within the new system). As noted earlier in this chapter, pay systems supporting FSAs might need to be implemented twice for the same population of service members, once under the legacy pay systems and again under any new service-specific systems (IPPS-A, AFIPPS, NP2).

The SMEs were, in general, unable to provide an estimate of the cost associated with implementing FSA support in the pay systems. The single ballpark estimate for a basic implementation that we obtained was $4 million for a single service's forthcoming pay and personnel system. So, we could multiply this by four to come up with an (admittedly crude) estimate of $16 million across all four services. In addition, if it was decided to upgrade the legacy DFAS DJMS to make FSA options available as quickly as possible, it might be reasonable to suppose an additional cost of between $4 million and $12 million, because this system currently supports the Army, Navy, Air Force, and Space Force. Thus, a very rough approximation of the total software implementation cost across both the legacy and new systems would be $20 million to $28 million, and potentially more given DoD's previous experience with pay and personnel software development efforts.

Our conclusion from the SME discussions is that the most substantial implementation cost would be that associated with upgrading the pay systems to support FSAs. The initial start-up costs associated with FSA administration and creating a training program for service members would likely be substantially smaller because of the availability of off-the-shelf materials and existing OPM materials.

Summary

In this chapter, we estimated potential savings to DoD and active-duty service members associated with instituting FSAs. We showed that members typically would benefit if they used FSAs, assuming they have eligible expenses. We consider a range of participation or take-up rates and believe that a 25-percent take-up rate, possibly higher for a DCFSA, is a reasonable first estimate. We also showed that, taking into account the costs we could quantify, DoD would also potentially experience savings in the long run if FSAs were implemented with the amount dependent on the take-up rates and the amounts that service members choose to contribute. Assuming a 25-percent take-up rate, the total tax savings to members would participate would be $51.2 million for an DCFSA and $33.1 million for an HCFSA under a $500 contribution assumption. For DoD, the respective savings in the first year and each subsequent year would be $29.8 million and $31.6 million for a DCFSA. For the HCFSA, DoD would experience a cost of $302,000 in the first year and savings of $6 million each subsequent year.

However, DoD would also bear costs that we were unable to quantify, specifically those costs associated with implementing changes to pay systems to support FSAs, creating FSA training programs for service members, and costs associated with setting up administration of the FSA benefit. These costs could be potentially large if implementation involved adapting both the legacy pay systems and the forthcoming service-specific systems, ranging up to $28 million or more. Furthermore, implementation could also result in delays in the rollout of the new IPPSs for the Army, Navy, and Air Force. In addition, DoD would bear ongoing overhead costs that we were not able to quantify.

CHAPTER SIX

Implementation Plan

This chapter outlines an implementation plan for HCFSA and DCFSA options for service members and their families. Here, we cover the following:

- the basic objective of the implementation plan and criteria for evaluating successfully meeting the objective
- a list of activities that will need to be carried out along with the key players
- a risk analysis covering the uncertainties associated with executing the plan
- a resource plan indicating what resources will need to be made available
- a provisional timeline and associated milestones.

Objective and Criteria

Objective: Allow all active-duty service members to participate in FSA options for health and dependent care in accordance with the congressional language in the FY 2021 NDAA, specifically "flexible spending account options that allow pre-tax payment of dependent care expenses, health and dental insurance premiums, and out-of-pocket health care expenses for members of the uniformed services and their family members." (U.S. House of Representatives Report 116-617, 2020)

Note that the objective is not simply to provide access to FSAs as they are currently implemented in much of the private sectors and for government civilians but to also provide a unique feature: the ability to use FSA funds to pay for health and dental insurance premiums. Under current law, FSA funds can only be used to pay for direct medical care services or goods, and not for insurance premiums. We will discuss this in more detail in this chapter.

Criterion for success: All active-duty service members have access to health and dependent care FSAs.

The criterion for success is broad in the sense that active-duty members of all services would need to be covered, and minimal in the sense that it only requires that active-duty members have access.[1] Additional criteria could include each service having a robust education and training program to support members in their choices, and also include ensuring

[1] We do not consider provision of FSAs to members of the reserve component in this report.

that every member in each of the services gets access simultaneously, because differences in timing would create inequities across services.

Roles and Responsibilities of Key Actors

The key actors involved in implementation include OSD, OPM (specifically staff supporting FSAFEDS and OPM's supporting contractor Health Equity), DFAS, and the proponents for the pay and personnel systems in each of the individual services. The key tasks they would need to undertake to bring about FSAs for active-duty members of the services and their family members would include

- deciding on the timing of when FSAs would be implemented
- seeing to it that enabling legislation for the novel FSA features comes before Congress
- implementing support for FSAs in both current and future pay and personnel systems
- creating education and training materials on FSAs for members of the services
- fielding the training in advance of the advent of the FSA options.

Table 6.1 shows provisional assignments of the roles and responsibilities to the different actors.

OSD would perform overall coordination, decide on the timing of implementation, and see to it that supporting legislation is drafted. The timing of implementation will affect the overall cost of implementation, as three of the services are transitioning from having their payroll services performed by DFAS to each having its own IPPS. These IPPSs are currently

TABLE 6.1
Roles and Responsibilities of Key Actors

Roles and Responsibilities	OSD	OPM	Health Equity	DFAS	Services
Coordination and timing decisions	X				X[a]
Facilitate enabling legislation	X				
Prepare education and training materials	X				X
Field education and training materials					X
Preparation and oversight of execution of FSA program		X			
Execution of FSA program			X		
Implement support for FSAs in current and future pay and personnel systems				X	X

NOTE: Assignment of roles and responsibilities is provisional.

[a] Services would support OSD in timing decisions because they are responsible for the release schedules for the new pay and personnel systems.

(as of June 2022) scheduled to be fully operational in FYs 2024 and 2025. If FSAs were to be implemented as soon as possible, then the supporting software might need to be implemented multiple times, once for the current pay systems and once again for the forthcoming IPPS. The alternative would be to delay implementation until fully operational versions of the new IPPS have been fielded, or mandate that the new IPPS include support for FSAs (and potentially delay when IPPS with full operational capability come online).

OSD Legislative Affairs would facilitate consideration of appropriate supporting legislation by Congress, specifically legislation that would allow active-duty service members to use FSA funds to pay for dental or health insurance premiums. In addition, and at the discretion of OSD, additional legislation might be supported that would allow FSAs to be stackable with employer subsidies.

DoD Office of Financial Readiness would take the lead in preparing education and training materials so that service members could take full advantage of the FSAs. It would need to coordinate with FSAFEDS and the individual services.

OPM would provide support via its FSAFEDS program staff. According to SME discussions, OPM staff would require a year to gear up to support FSAs as currently designed for all members of the services and might require additional time to support FSAs that allow for payment of health and dental insurance premiums. This period would include developing the processes and contractual arrangements required to administer the plan. Health Equity, which administers the FSAFEDS program on behalf of OPM and takes care of day-to-day operations, would similarly require a year or longer to build the capacity needed to support all members of the services and their families.

DFAS would need to implement support for FSAs in all their current pay systems if FSAs are to be implemented before the individual service pay and personnel systems come online. DFAS currently oversees the DJMS, which provides payroll support for the Army, Navy, Air Force and Space Force. DFAS also oversees the Marine Corps Total Force System (MCTFS). If FSAs are to be implemented only after all the individual service pay and personnel systems come online, then DFAS would only need to implement support for FSAs in the MCTFS.

The **services** developing new pay and personnel systems would need to implement support for FSAs. In addition, the services would need to assist OSD in making timing decisions for the provision of FSAs, taking into consideration the development schedule for when the new systems will have full operational capability. The services could also assist OSD in the development of education and training materials and would have responsibility for fielding these materials.

Risk Analysis

The primary risks associated with implementing FSAs are associated with enabling legislation and development of supporting pay system software. It is uncertain if and when enabling legislation could be passed that would allow active-duty service members to pay for dental and health insurance using FSA funds. It is also uncertain whether and when auxiliary legislation could be passed that would allow active duty service members to "stack" DCFSA funds and service subsidies. The Department of the Treasury could object to either or both of these provisions unless there were offsets to the loss of revenue that would be induced by any proposed legislation. The uncertainty surrounding legislation would lead to uncertainty for the other the key actors, OPM, DFAS, and the services. They may be unable or unwilling to act until any residual legislative uncertainty has been resolved.

Development of software to support FSAs in the legacy and new pay systems could prove to be a formidable challenge, perhaps most comparable with the effort to support the launch of the TSP for military members. The development and maintenance of pay and personnel software can often prove to be more difficult than initially anticipated, as can be seen from the efforts that have been delayed or abandoned in the past (e.g., the Defense Integrated Military Human Resources System). That being said, FSAFEDS has experience in providing a functional application support interface to the federal agencies with which it works, and DFAS and the services would have these other implementations that could serve as a guide to providing the needed software support. The overall timing of the rollout of FSAs will determine whether support would need to be developed within DJMS. In any case ,the development timeline for the supporting software for FSAs could prove to be a key source of uncertainty.

Resource Plan

As we noted in Chapter Five, our conclusion from the SME discussions is that the most substantial implementation cost would be the cost associated with upgrading the pay systems to support the FSA options. We projected from a single service's ballpark estimate of $4 million to provide a minimal level of support for FSAs in one of the new pay and personnel systems to suppose that the implementation cost to upgrade the four new or continuing pay systems to be $16 million, while the implementation cost to upgrade the legacy and the new pay systems (and thus potentially provide FSAs before the new pay systems reach full operational capability) would be in the range of $20 million to $28 million or possibly much more given previous experience. The initial start-up costs associated with FSA administration and creating a training program for service members would likely be substantially smaller.

Timeline and Milestones

Were DoD to decide to implement FSAs immediately without waiting for the services to complete their upgrade of IPPS, it would likely take well over a year to accomplish. This is due to several factors. First, implementing flexible account options that would allow pre-tax payment of health and dental insurance premiums would require enabling legislation. Second, three of the services are currently implementing new pay and personnel systems and would require time to develop support for FSAs once their new systems are fielded in FYs 2024 and 2025. Third, OPM and its supporting contractor Health Equity would require a year to prepare for providing FSA accounts, and possibly more time if the FSA accounts have novel features, such as allowing payment of health and dental insurance premiums. Fourth, time would be required to develop training materials for service members. Fortunately, some of these activities could be carried out in parallel, as they are not dependent on each other.

This is illustrated by the example schedule shown in Figure 6.1. This schedule assumes that OSD decides to implement FSAs as quickly as possible, and thus has DFAS implement support for FSAs in DJMS. Initially, OSD facilitates enabling legislation to allow members of the services to pay for dental and health insurance premiums using FSAs. The example schedule allows one year for this activity, however there is a possibility that it could take longer, and project managers will need to take this risk into account. Subsequent to the legislation being passed, OPM and their supporting contractor could prepare to support the FSA program. Simultaneously, DFAS could initiate work to implement support of FSAs in the systems it supports; although the example schedule allocates one year for this activity, one of the risks that would need to be managed is that this could take considerably longer. (If DJMS support for FSAs where to take two or more years to accomplish then implementation of FSAs would slip to 2026, by which time the new service pay and personnel systems are scheduled to be operational.) Also, the DoD Office of Financial Readiness could simultaneously begin work on preparing the education and training materials, later handing these materials off to the services. Thus, OPM, DFAS, OSD, and the services would work in parallel toward the initial offering of FSAs to service members, which in this scenario would happen in 2025. OPM would oversee the execution of the FSA program from 2025 into the indefinite future. The Army, Navy, and Air Force would field systems with full operational capability and support for FSAs by 2026, and they would transition from DFAS support at that time.

FSAs for service members could be implemented faster—perhaps up to a year faster—if enabling legislation to support such special features as payment of insurance premiums was not required. If the "OSD-facilitate enabling legislation" block was removed from the scenario shown in Figure 6.1, the succeeding activities could all be moved to the left by year, with "OPM, contractor-preparation of FSA programs," "DFAS-support FSAs in current pay systems," and "OSD, services-prepare education and training materials" all being able to start

FIGURE 6.1
Example Schedule for Implementation

as soon as resources could be allocated. Thus, barring any unexpected delays (due to, say, software development difficulties), FSAs without special features could potentially be made available by 2024 instead of 2025.

An alternative scenario, shown in Figure 6.2, would be to avoid the cost of implementing support for FSAs in the legacy DJMS pay system, and would instead only implement FSAs on the continuing and new pay systems. This scenario would allow some slack in the timing of support activities carried out by OPM and its contractor, as well as the training and education activities carried out by OSD and the services—therefore, these tasks have been moved to the figure's right so that they occur in 2025 for the most part but could occur earlier. Under this scenario, FSAs would be rolled out when all the new service pay and personnel systems reach full operational capability, here assumed to be by the beginning of 2026. OPM would provide support for the FSA program from 2026 on into the indefinite future.

The assumption that the new systems are available at the beginning of 2026 reflects the current (June 2022) schedule for the new IPPSs across all the services, however these schedules are subject to change. Major software development projects are subject to both time and budget risk, and typically take longer and cost more than initially anticipated. Requiring that the new systems provide support for FSAs could delay their delivery. Thus, the rollout of FSAs could inadvertently be delayed by making the scheduling dependent on when the new pay and personnel systems reach full operational capability.

FIGURE 6.2
Alternative Example Schedule for Implementation

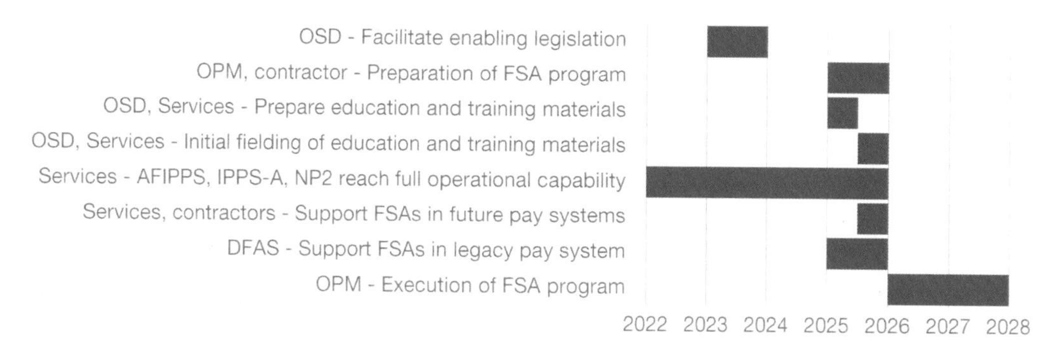

NOTE: The figure assumes that OPM, OSD, and service preparation tasks are carried out in 2025 for the most part, but they could occur earlier.

Summary

In this chapter, we described a basic implementation plan for FSAs for service members and their families. We outlined the objective and criteria for success, described the key actors and activities they would need to undertake, gave a basic risk analysis (identifying enabling legislation and software development as the key sources of risk), gave a provisional resource plan, and provided two example timelines. Implementation of FSAs is possible and could happen as early as 2025 under the more aggressive example schedule we have presented—if all tasks can be executed within the time planned. If there are delays due to difficulties in realizing the enabling legislation, or due to software development problems, or due to OPM (or their supporting contractor) needing additional time to support an FSA program with novel features, then this target date could slip to 2026 or even later. The less aggressive schedule is also subject to risk, as it makes implementation of FSAs dependent on the delivery of three large and complex software products, AFIPPS, IPPS-A, and NP2, and delays in any one of these systems could potentially disrupt the scheduled time when FSAs would become available to all active-duty service members.

CHAPTER SEVEN

Discussion and Conclusions on the Feasibility and Advisability of FSAs for Service Members

This chapter draws from the results of the previous chapters to discuss the feasibility of implementing FSA options that would allow pre-tax payment of dependent care expenses and health-related expenses for members and their families. In considering feasibility, Congress directed an analysis to evaluate the financial advantages or disadvantages for members and their families, focusing on tax incentives and identify administrative or legislative barriers to implementing these options. The net effect of an FSA on a member's tax liability is not predictable *a priori*. On the one hand, members' income tax liability may rise, fall, or stay the same because of the complex interactions between FSA contributions and other relevant tax credits. On the other hand, payroll taxes fall for both members and DoD because less income is subject to Social Security and Medicare taxes. The net effect on DoD is also not predictable *a priori* because despite the drop in payroll taxes, DoD must pay for the ongoing administration of the FSA options, incurring a cost. In addition to the analysis directed by Congress, we also developed an implementation plan should DoD move forward with these options, at the request of OSD.

We find that across all active service members who would likely be eligible to participate in the FSA options, under a wide range of assumptions about the participation rate, the aggregate net benefit to active service members and to DoD would be positive.[1] Although the overall effect is estimated to be positive, most active members would likely not participate because they would either not have eligible expenses or their expenses would be relatively small. In the case of a DCFSA, members with children in off-base subsidized child care programs would be unlikely to have eligible expenses, unlike those with children in on-base care, such as in a CDC. Yet, spaces in on-base care facilities are supply constrained so not every military family who would prefer to receive on-base care (and fully benefit from the DCFSA) would be able to do so. In the case of an HCFSA, active members receive health care at little

[1] When a $500 HCFSA contribution is assumed, we estimate that DoD would experience a cost to offering an HCFSA in the first year. However, this cost is small relative to the savings estimated in each subsequent year, meaning that the DoD would experience a net savings from offering an HCFSA after two years.

or no out-of-pocket costs to them while most of their families, 83 percent, are covered by TRICARE Prime with average annual out-of-pocket costs of about $100 or less. The remaining 17 percent of active families are covered by TRICARE Select with average annual out-of-pocket costs of about $500.

Our computation of net benefit to DoD does not include implementation or ongoing overhead costs. For DoD, the cost to implementing FSA options could be sizable, albeit currently unknowable. Cost could be large because of the costs of upgrading the pay systems to support FSAs, not just the current systems but the IPPSs that are being rolled out by each service over the next few years.

The overall implication is that adding FSA options for service members and their families would impart an overall net benefit to service members who have eligible expenses and would produce a cost savings to DoD. However, available data indicate many members would not benefit from an FSA option in the case of dependent care or, in the case of health care, their average out-of-pocket expenses would be small, less than $100 annually. Furthermore, the potential inequity between those receiving on-base versus off-base child care in terms of access to the benefits of an FSA option could increase concerns regarding the limited availability and the existence of waitlists for on-base care. Improved data on usage of unsubsidized off-base child care and on out-of-pocket health care expenses outside TRICARE would improve the estimates of the net advantage to service members and their families.[2] Although the FSA options would produce a cost savings to DoD, the benefits to DoD of reduced payroll costs are an intergovernmental transfer, while the administrative costs are a real cost to the taxpayer. We were unable to provide a specific estimate of implementation costs beyond a rough estimate of up to $28 million or more, but our SME discussions suggest that one way to reduce costs would be to wait to implement the FSA options until all services have full transitioned to their IPPSs, to avoid duplicating implementation costs under the legacy systems.

Changes in Legislation

Two legislative changes that could potentially expand the benefits of an FSA option to more service members and their families or that would be required to implement FSA option for health care as articulated by Congress.

First, a prominent challenge to implementation of FSAs as envisaged in the NDAA 2021 language is that current regulations do not allow for health FSA funds to be applied to health and dental insurance premiums. HCFSA funds can be spent only on medical care. The Internal Revenue Code states that *medical care* means amounts paid "for the diagnosis, cure, mitigation, treatment or prevention of disease, or for the purpose of affecting any structure or function of the body" (U.S. Code, Title 26, Section 213[d], 2012). As health and dental insur-

[2] Future DoD surveys should consider adding questions related to these issues.

ance premiums are not a form of direct care, they cannot be paid for out of an FSA.[3] Thus, enabling legislation would need to be passed to allow for payment of insurance premiums from FSAs by military personnel. Active-duty members and their families have no premium or enrollment fee for TRICARE Prime or Select, but they pay a premium for dental care. For coverage of a single individual, annual premiums are $139.80 and are $363.36 for families in 2022. Allowing an HCFSA to cover these premiums could increase participation in the HCFSA.

Second, active members are more likely to benefit from a DCFSA if their dependent care is provided on-base, a form of care that is supply constrained, or provided off base from an unsubsidized provider. Although military families may find certified off-base providers where expenses are eligible for fee assistance, this option would most likely eliminate the advantage of having a DCFSA because any combination of child care subsidy and DCFSA contributions cannot exceed the maximum DCFSA contribution limit of $5,000. For example, a member with TFI of $55,000 would pay $100 weekly for community-based care, or $5,200 annually. Suppose the cost of child care was $20,000 annually. Under the fee-assistance program, the military would have provided the subsidy up to the cap of $1,500 monthly, or $18,000 annually, less the fees paid by a family based on its TFI, for a total subsidy of $12,800. A family would also have $2,000 in additional out-of-pocket costs (equal to $20,000 – $5,200 – $12,800). This $12,800 subsidy far exceeds the contribution limit of $5,000 for a DCFSA, and a family would have $0 in allowable contributions to a DCFSA. Also not allowable would be any child care fees above the $20,000 figure in this example. For example, if, instead of $20,000, the provider charged $24,000 annually, the member would be responsible for $11,200 = $5,200 + $4,000 (where $4,000 = $24,000 – $20,000). The additional $4,000 would also not be allowable.

In short, implementing a DCFSA would create an inequity between members who receive on-base child care in CDCs and those who receive subsidized community-based care. Concerns about this equity in accessing the benefits of a DCFSA could potentially be addressed by eliminating the offset for the employer subsidy of child care. However, eliminating the offset would require legislation to make an employer child care subsidy and eligible DCFSA expenses stackable. Allowing them to be stackable would likely increase participation in the DCFSA.

Finally, FSAs for service members could be implemented faster—perhaps up a year faster—if enabling legislation to support special features, such as payment of insurance premiums, was not required. Enabling legislation for special features lies on the critical path for implementation of FSAs because it is the precursor to the development of any processes, procedures, and software to support fielding of FSAs—these cannot be developed without knowing the particulars of the legislation.

[3] As discussed in Chapter Two, FSAs for medical care can cover direct expenditures eye and dental care including orthodontia, eye exams, glasses, and so forth.

Summary

Introducing an FSA option could impart a tax benefit for some members and their families and save personnel costs for DoD, although the start-up costs of updating the legacy and new pay systems could be substantial. However, many military families would not have eligible DCFSA expenses or could have few eligible HCFSA expenses under current law, unless they use the HCFSA to cover expenses outside TRICARE, such as for over-the-counter medication and supplies. Furthermore, the DCFSA introduces an inequity between those who already benefit from having access to on-base care and those who access care off base from subsidized providers. Legislation that addresses these issues could increase participation in the FSA options.

Abbreviations

AGI	adjusted gross income
AFIPPS	Air Force Integrated Pay and Personnel System
CDC	child development center
CDCTC	Child and Dependent Care Tax Credit
CONUS	continental United States
DCFSA	dependent care flexible spending account
DEERS	Defense Enrollment Eligibility Reporting System
DFAS	Defense Finance and Accounting Service
DJMS	Defense Joint Military Pay System
DMDC	Defense Manpower Data Center
DoD	U.S. Department of Defense
EITC	Earned Income Tax Credit
FCC	family child care
FICA	Federal Income Contribution Act
FSA	flexible spending account
FSAFEDS	Federal Flexible Spending Account Program
FY	fiscal year
HCFSA	health care flexible spending account
HMO	health maintenance organization
HRA	health reimbursement arrangement
HSA	health savings account
IPPS	Integrated Pay and Personnel System
IPPS-A	Integrated Pay and Personnel System–Army
IRS	Internal Revenue Service
MCTFS	Marine Corps Total Force System
NBER	National Bureau of Economic Research
NDAA	National Defense Authorization Act
NP2	Navy Personnel and Pay System
OPM	Office of Personnel Management
OSD	Office of the Secretary of Defense
PCM	primary care manager
PITE	Point-in-Time Extract
PPO	preferred provider organization
SAC	school-aged care

SME	subject-matter expert
TFI	total family income
TSP	Thrift Savings Plan

References

Bureau of Labor Statistics, "Consumer Expenditure Surveys, Table R-1. All Consumer Units: Annual Detailed Expenditure Means, Standard Errors, Coefficients of Variation, and Weekly or Quarterly Percents Reporting," Washington, D.C., 2020. As of May 10, 2022: https://www.bls.gov/cex/tables/calendar-year/mean/cu-all-detail-2020.pdf

Burke, Jeremy, and Amalia R. Miller, "The Effects of Job Relocation on Spousal Careers: Evidence from Military Change of Station Moves," *Economic Inquiry*, Vol. 56, No. 2, 2018, pp. 1261–1277.

Feenberg, Daniel Richard, and Elizabeth Coutts, "An Introduction to the TAXSIM Model," *Journal of Policy Analysis and Management*, Vol. 12, No. 1, 1993, pp. 189–194.

Child Care Aware of America, "Fee Assistance," webpage, undated-a. As of May 9, 2022: https://www.childcareaware.org/fee-assistancerespite/feeassistancerespiteproviders/feeassistance/

Child Care Aware of America, "Find and Afford Quality Child Care in the Community," webpage, undated-b. As of May 9, 2022: https://www.childcareaware.org/wp-content/uploads/2022/04/MCCYN-PLUS-Fact-Sheet.pdf

Child Care Aware of America "FY 22 Child Care Fee Assistance Total Family Income Categories, DoD Parent Fees, High Cost Installations, and Provider Caps," 2021. As of May 10, 2022: https://www.childcareaware.org/wp-content/uploads/2021/09/FY22-NAFMC-Fee-Categories-and-Parent-Fees-1.pdf

Child Care Aware of America, *Demanding Change: Repairing our Child Care System*, Arlington, Va., March 2022a. As of May 10, 2022: https://info.childcareaware.org/hubfs/2022-03-FallReport-FINAL%20(1).pdf?utm_campaign=Budget%20Reconciliation%20Fall%202021&utm_source=website&utm_content=22_demandingchange_pdf_update332022

Child Care Aware of America, *Demanding Change: Repairing our Child Care System, Appendices*, Arlington, Va., March 2022b. As of May 10, 2022: https://info.childcareaware.org/hubfs/Demanding%20Change%20Appendices.pdf?utm_campaign=Budget%20Reconciliation%20Fall%202021&utm_source=website&utm_content=22_demandingchange_append

Crandall-Hollick, Margot L., *The Child Tax Credit: Current Law*, Washington, D.C.: Congressional Research Service, R41873, updated May 15, 2018. As of May 9, 2022: https://crsreports.congress.gov/product/pdf/R/R41873/20#:~:text=Eligible%20families%20can%20claim%

Crandall-Hollick, Margot L., "The Child and Dependent Tax Credit (CDCTC): Temporary Expansion for 2-21 Under the American Rescue Plan Act of 2021 (ARPA; P.L. 117-2)," Congressional Research Service, Insight No. IN11645, updated May 10, 2021. As of May 9, 2022: https://crsreports.congress.gov/product/pdf/IN/IN11645

Crandall-Hollick, Margot L., Gene Falk, and Conor F. Boyle, *The Earned Income Tax Credit (EITC): How It Works and Who Receives It*, Washington, D.C.: Congressional Research Service, January 12, 2021a. As of May 9, 2022: https://crsreports.congress.gov/product/pdf/R/R43805#_Toc61454427

Crandall-Hollick, Margot L., Jameson A. Carter, and Conor F. Boyle, *The Child Tax Credit: The Impact of the American Rescue Plan Act (ARPA P.L. 117-2) Expansion on Income and Poverty*, Washington, D.C.: Congressional Research Service, R46839, July 13, 2021. As of May 9, 2022: https://crsreports.congress.gov/product/pdf/R/R46839#:~:text=CRS%20estimates%20that%20as%20a,among%20the%20lowest%2Dincome%20families

Dicken, John, *Health Savings Accounts: Participation Increased and Was More Common Among Individuals with Higher Incomes*, Washington, D.C., U.S. General Accountability Office, GAO-08-474R, 2008. As of May 6, 2022:
https://www.gao.gov/assets/gao-08-474r.pdf

Federal Register, "Federal Retirement Thrift Investment Board," webpage, undated. As of May 10, 2022:
https://www.federalregister.gov/agencies/federal-retirement-thrift-investment-board

Garren, Gwen, "Get a Closer Look At the Earned Income Tax Credit (EITC), a Federal Tax Credit That Gives American Workers and Families a Financial Boost," Internal Revenue Service, February 17, 2022. As of May 9, 2022:
https://www.irs.gov/about-irs/a-closer-look-at-the-earned-income-tax-credit

Hung, Man, Sharon Su, Eric S. Hon, Edgar Tilley, Alex Macdonald, Evelyn Lauren, Glen Roberson, and Martin S. Lipsky, "Examination of Orthodontic Expenditures and Trends in the United States from 1996 to 2016: Disparities Across Demographics and Insurance Payers," *BMC Oral Health*, Vol. 21, No. 268, May 2021. As of June 15, 2022:
https://www.ncbi.nlm.nih.gov/pmc/articles/PMC8130155/

Internal Revenue Service, "Notice 2013-17, Modification of "Use-or-Lose" Rule For Health Flexible Spending Arrangements (FSAs) and Clarification Regarding 2013-2014 Non-Calendar Year Salary Reduction Elections Under § 125 Cafeteria Plans," webpage, 2013. As of May 8, 2022:
https://www.irs.gov/pub/irs-drop/n-13-71.pdf

Internal Revenue Service, "Notice 2021-15, Additional Relief for Coronavirus Disease (COVID-19) Under § 125 Cafeteria Plans," webpage, 2021a. As of May 8, 2022:
https://www.irs.gov/pub/irs-drop/n-21-15.pdf

Internal Revenue Service, *26 CFR 601.602: Tax Forms and Instructions, Rev. Proc. 2021-45: Inflation-Adjusted Items for 2022 for Various Provisions of the Internal Revenue Code of 1986 (Code)*, Washington, D.C., November 10, 2021b. As of May 9, 2022:
https://www.irs.gov/pub/irs-drop/rp-21-45.pdf

Internal Revenue Service, *Publication 503: Child and Dependent Care Expenses*, Washington, D.C., December 20, 2021c. As of May 8, 2022:
https://www.irs.gov/pub/irs-pdf/p503.pdf

Internal Revenue Service, *Publication 969: Health Savings Accounts and Other Tax-Favored Health Plans*, Washington, D.C., January 6, 2022a. As of May 8, 2022:
https://www.irs.gov/pub/irs-pdf/p969.pdf

Internal Revenue Service, *Publication 502: Medical and Dental Expenses (Including the Health Coverage Tax Credit)*, Washington, D.C., January 11, 2022b. As of May 8, 2022:
https://www.irs.gov/pub/irs-pdf/p502.pdf

Internal Revenue Service, "FAQs for Government Entities Regarding Cafeteria Plans," webpage, updated January 20, 2022c. As of May 8, 2022:
https://www.irs.gov/government-entities/federal-state-local-governments/faqs-for-government-entities-regarding-cafeteria-plans

Internal Revenue Service, "Military and Clergy Rules for the Earned Income Tax Credit," webpage, January 27, 2022d. As of May 9, 2022:
https://www.irs.gov/credits-deductions/individuals/earned-income-tax-credit/military-and-clergy-rules-for-the-earned-income-tax-credit#How%20to%20Include%20Nontaxable%20Pay%20When%20You%20Claim%20the%20EITC

Internal Revenue Service, *Publication 15-B: Employer's Tax Guide to Fringe Benefits*, Washington, D.C., January 31, 2022e. As of May 8, 2022:
https://www.irs.gov/pub/irs-pdf/p15b.pdf

Internal Revenue Service, "Who Qualifies for the Earned Income Tax Credit (EITC)," webpage, updated March 14, 2022f. As of May 9, 2022:
https://www.irs.gov/credits-deductions/individuals/earned-income-tax-credit/who-qualifies-for-the-earned-income-tax-credit-eitc

Internal Revenue Service, "FS 2022-28: IRS Updates Tax Year 2021/Filing Season 2022 Child Tax Credit Frequently Asked Questions, Information to Help Taxpayers Prepare Their 2021 Returns," fact sheet, April 2022g. As of May 9, 2022:
https://www.irs.gov/pub/taxpros/fs-2022-28.pdf

IRS—*See* Internal Revenue Service.

Kamarck, Kristy N., *Military Child Development Program: Background and Issues*, Washington, D.C.: Congressional Research Service, R45288, March 19, 2020. As of May 9, 2022:
https://crsreports.congress.gov/product/pdf/R/R45288/7

Lo Sasso, Anthony, Lorens Helmchen, and Robert Kaestmer, "The Effects of Consumer-Directed Health Plans on Health Care Spending," *Journal of Risk and Insurance*, Vol. 77, No. 1, Special Issue on Health Insurance, 2010, pp. 85–103. As of June 10, 2022:
https://www.jstor.org/stable/20685291

Lorman Education Services, "Flexible Spending Accounts," webpage, undated. As of May 23, 2022:
https://www.lorman.com/training/benefits/flexible-spending-accounts

Military Health System, "Rates and Reimbursement. TRICARE Prime and Select Calendar Year 2022 Out of Pocket Costs: Active Duty Family Members," webpage, undated. As of May 8, 2022:
https://health.mil/Military-Health-Topics/Access-Cost-Quality-and-Safety/TRICARE-Health-Plan/Rates-and-Reimbursement

National Bureau of Economic Research, "TAXSIM Related Files at the NBER," webpage, undated. As of July 28, 2022:
http://taxsim.nber.org/

NBER—*See* National Bureau of Economic Research.

Office of the Assistant Secretary of Defense for Health Affairs, Defense Health Agency, *Evaluation of TRICARE Program: Fiscal Year 2021 Report to Congress*, Washington, D.C., February 26, 2021. As of May 10, 2022:
https://health.mil/Military-Health-Topics/Access-Cost-Quality-and-Safety/Health-Care-Program-Evaluation/Annual-Evaluation-of-the-TRICARE-Program

Office of the Assistant Secretary of Defense for Manpower and Reserve Affairs, "Subject: DoD Child Development Program Fees for School Year 2021-21," memorandum for Assistant Secretary of The Army (Manpower And Reserve Affairs) Assistant Secretary Of The Navy (Manpower And Reserve Affairs) Assistant Secretary Of The Air Force (Manpower And Reserve Affairs) Director, Defense Logistics Agency, Washington, D.C., May 5, 2021.

Office of People Analytics, U.S. Department of Defense, "2019 Survey of Active Duty Spouses: Infographic on Spouse Education and Employment," webpage, May 1, 2020. As of March 1, 2022:
https://www.opa.mil/research-analysis/spouse-family/military-spouse-survey-survey-reports-briefings

Office of Personnel Management, "FSAFEDS," homepage, undated. As of May 23, 2022:
https://www.fsafeds.com

Office of Personnel Management, "OPM Benefits Administration Letter Number 18-801, The Federal Flexible Spending Account Program (FSAFEDS): 2018 Administrative Fees," Washington, D.C., February 9, 2018. As of April 27, 2022:
https://www.opm.gov/retirement-services/publications-forms/benefits-administration-letters/2018/18-801.pdf

Office of Personnel Management, "OPM Benefits Administration Letter Number 20-801, The Federal Flexible Spending Account Program (FSAFEDS): 2020 Administrative Fees," Washington, D.C., April 27, 2020. As of April 27, 2022:
https://www.opm.gov/retirement-services/publications-forms/benefits-administration-letters/2020/20-801.pdf

Office of Personnel Management, "OPM Benefits Administration Letter Number 21-801, Announcing New Third-Party Administrator and Billing Requirements for The Flexible Spending Account Program (FSAFEDS) Effective January 1, 2021," Washington, D.C., January 12, 2021. As of April 27, 2022:
https://www.opm.gov/retirement-services/publications-forms/benefits-administration-letters/2021/21-801.pdf

Tax Policy Center, "Briefing Book: Key Elements of the U.S. Tax System. What Is the Difference Between Refundable and Nonrefundable Credits?" webpage, updated May 2020. As of May 8, 2022:
https://www.taxpolicycenter.org/briefing-book/what-difference-between-refundable-and-nonrefundable-credits#:~:text=REFUNDABLE%20VERSUS%20NONREFUNDABLE%20TAX%20CREDITS,liability%20is%20refunded%20to%20taxpayers

Tax Policy Center, "EITC Parameters," webpage, updated March 19, 2021. As of May 10, 2022:
https://www.taxpolicycenter.org/statistics/eitc-parameters

TRICARE, "Dental Costs," webpage, updated September 28, 2021. As of May 8, 2022:
https://www.tricare.mil/Costs/DentalCosts

TRICARE, "Select Enrollment Fees," webpage, updated February 15, 2022a. As of May 8, 2022:
https://www.tricare.mil/Plans/Enroll/Select/EnrollmentFees

TRICARE, "TRICARE Prime," webpage, updated February 15, 2022b. As of May 8, 2022:
https://www.tricare.mil/Plans/HealthPlans/Prime

TRICARE, "TRICARE Dental Program Costs Cost Shares" webpage, updated May 2, 2022c. As of May 8, 2022:
https://www.tricare.mil/Costs/DentalCosts/TDP/CostShares/

TRICARE, "TRICARE Dental Program Costs Monthly Premiums," webpage, updated May 2, 2022d. As of May 8, 2022:
https://www.tricare.mil/Costs/DentalCosts/TDP/Premiums

TRICARE, "TRICARE Dental Program Costs Plan Maximums," webpage, updated May 2, 2022e. As of May 8, 2022:
https://www.tricare.mil/Costs/DentalCosts/TDP/Maximums

U.S. Air Force, "AFIPPS," webpage, undated. As of May 23, 2022:
https://www.jber.jb.mil/Services-Resources/Personnel/AFIPPS/

U.S. Army, "IPPS-A," webpage, undated. As of May 23, 2022:
https://ipps-a.army.mil

U.S. Code, Title 26, Internal Revenue Code; Section 213, Medical, Dental, etc., Expenses; (d) Definitions, 2012.

U.S. Department of Defense, "Military Compensation: Monthly Basic Pay Table, Effective 1 January 2020," webpage, January 2020. As of March 1, 2022:
https://militarypay.defense.gov/Portals/3/Documents/ActiveDutyTables/2020%20Military%20Basic%20Pay%20Table.pdf

U.S. Department of Health and Human Services, Administration for Children and Families, "45 CFR Part 98, Child Care and Development Fund (CCDF) Program," *Federal Register*, Vol. 81, No. 190, September 30, 2016. As of May 10, 2022:
https://www.govinfo.gov/content/pkg/FR-2016-09-30/pdf/2016-22986.pdf

U.S. Government Accountability Office, "Military Child Care: Potential Costs and Impacts of Expanding Off-Base Child Care Assistance for Children of Deceased Servicemembers," report to U.S. Senate and U.S. House of Representatives Committees on Armed Services, Washington, D.C., GAO-22105186, December 14, 2021. As of August 8, 2022:
https://www.gao.gov/assets/gao-22-105186.pdf

U.S. House of Representatives Report 116-617, *William M. (Mac) Thornberry National Defense Authorization Act for Fiscal Year 2021, Section 750, Plan for Evaluation of Flexible Spending Account Options for Members of the Uniformed Services and Their Families*, conference report to accompany House Resolution 6395, Washington, D.C.: U.S. Government Publishing Office, December 3, 2020. As of July 27, 2022:
https://www.congress.gov/congressional-report/116th-congress/house-report/617/1

U.S. Navy, "NP2," webpage, undated. As of May 23, 2022:
https://my.navy.mil/np2.html